273
Topics in Current Chemistry

Topics in Current Chemistry
Recently Published and Forthcoming Volumes

STM and AFM Studies on (Bio)molecular Systems: Unravelling the Nanoworld
Volume Editor: Samori, P.
Vol. 285, 2008

Amplification of Chirality
Volume Editor: Soai, K.
Vol. 284, 2008

Anthracycline Chemistry and Biology II
Mode of Action, Clinical Aspects and New Drugs
Volume Editor: Krohn, K.
Vol. 283, 2008

Anthracycline Chemistry and Biology I
Biological Occurence and Biosynthesis, Synthesis and Chemistry
Volume Editor: Krohn, K.
Vol. 282, 2008

Photochemistry and Photophysics of Coordination Compounds II
Volume Editors: Balzani, V., Campagna, S.
Vol. 281, 2007

Photochemistry and Photophysics of Coordination Compounds I
Volume Editors: Balzani, V., Campagna, S.
Vol. 280, 2007

Metal Catalyzed Reductive C–C Bond Formation
A Departure from Preformed Organometallic Reagents
Volume Editor: Krische, M. J.
Vol. 279, 2007

Combinatorial Chemistry on Solid Supports
Volume Editor: Bräse, S.
Vol. 278, 2007

Creative Chemical Sensor Systems
Volume Editor: Schrader, T.
Vol. 277, 2007

In situ NMR Methods in Catalysis
Volume Editors: Bargon, J., Kuhn, L. T.
Vol. 276, 2007

Sulfur-Mediated Rearrangements II
Volume Editor: Schaumann, E.
Vol. 275, 2007

Sulfur-Mediated Rearrangements I
Volume Editor: Schaumann, E.
Vol. 274, 2007

Bioactive Conformation II
Volume Editor: Peters, T.
Vol. 273, 2007

Bioactive Conformation I
Volume Editor: Peters, T.
Vol. 272, 2007

Biomineralization II
Mineralization Using Synthetic Polymers and Templates
Volume Editor: Naka, K.
Vol. 271, 2007

Biomineralization I
Crystallization and Self-Organization Process
Volume Editor: Naka, K.
Vol. 270, 2007

Bioactive Conformation II

Volume Editor: Thomas Peters

With contributions by

S. Afonin · A. S. Arseniev · T. Arvinte · J. L. Asensio
A. Bastida · M. J. J. Blommers · E. V. Bocharov · V. Dötsch
U. H. N. Dürr · M. Geiser · W. Jahnke · V. Jayalakshmi
J. Jiménez-Barbero · M. A. Johnson · N. R. Krishna · K. V. Pavlov
T. Peters · B. M. Pinto · C. Rademacher · P. Ramage · J. Salgado
H. Sparrer · A. Strauss · A. S. Ulrich ·P. Wadhwani

Springer

The series *Topics in Current Chemistry* presents critical reviews of the present and future trends in modern chemical research. The scope of coverage includes all areas of chemical science including the interfaces with related disciplines such as biology, medicine and materials science. The goal of each thematic volume is to give the nonspecialist reader, whether at the university or in industry, a comprehensive overview of an area where new insights are emerging that are of interest to a larger scientific audience.
As a rule, contributions are specially commissioned. The editors and publishers will, however, always be pleased to receive suggestions and supplementary information. Papers are accepted for *Topics in Current Chemistry* in English.
In references *Topics in Current Chemistry* is abbreviated Top Curr Chem and is cited as a journal.
Visit the TCC content at springerlink.com

ISBN 978-3-642-44747-1 ISBN 978-3-540-49080-7 (eBook)
DOI 10.1007/978-3-540-49080-7

Topics in Current Chemistry ISSN 0340-1022

Softcover re-print of the Hardcover 1st edition 2008

Cover design: WMXDesign GmbH, Heidelberg
Typesetting and Production: le-tex publishing services oHG, Leipzig

Printed on acid-free paper

9 8 7 6 5 4 3 2 1 0

springer.com

Volume Editor

Prof. Dr. Thomas Peters

Universität Lübeck
Inst. Chemie
Ratzeburger Allee 160
23538 Lübeck
thomas.peters@chemie.uni-luebeck.de

Editorial Board

Topics in Current Chemistry
Also Available Electronically

For all customers who have a standing order to Topics in Current Chemistry, we offer the electronic version via SpringerLink free of charge. Please contact your librarian who can receive a password or free access to the full articles by registering at:

springerlink.com

If you do not have a subscription, you can still view the tables of contents of the volumes and the abstract of each article by going to the SpringerLink Homepage, clicking on "Browse by Online Libraries", then "Chemical Sciences", and finally choose Topics in Current Chemistry.

You will find information about the

- Editorial Board
- Aims and Scope
- Instructions for Authors
- Sample Contribution

at springer.com using the search function.

Color figures are published in full color within the electronic version on SpringerLink.

Preface

Specific binding of a ligand to a receptor is a key step in a variety of biological processes, such as immune reactions, enzyme cascades, or intracellular transport processes. The ligand–receptor terminology implies that the receptor molecule is significantly larger than the ligand, and the term "bioactive conformation" usually characterizes the conformation of a ligand when it is bound to a receptor. In a more general sense, bioactive conformation applies to any molecule in a biologically relevant bound state regardless of size considerations. Most of the contributions to this book address ligands that are much smaller than their receptors.

X-ray crystallography and high resolution NMR spectroscopy are the two main experimental techniques used to study bioactive conformations. Therefore, the two volumes of this book cover approaches that use either of the two techniques, or a combination thereof. The combination of X-ray crystallography and NMR spectroscopy is particularly useful when a crystal structure of a receptor protein, but not of the receptor protein–ligand complex, is available. A number of experimental techniques to analyze the bioactive conformation of a ligand with NMR are based on the observation of the resonance signals of the free ligand that is in exchange with the bound ligand. Several chapters focus on such approaches that range from "classical" transferred NOE experiments, to transferred dipolar couplings, to STD (saturation transfer difference) NMR techniques. In cases where tight binding in the sub-nanomolar range prevents the analysis of the bioactive conformation via free ligand signals, the ligand–protein complex has to be analyzed with protein NMR-based techniques or by crystallography. Since this area has been the subject of many reviews and monographs it will not be covered here in particular detail. As a unifying theme, all contributions target the question of how molecular recognition of biologically active molecules is achieved on the atomic scale. Depending on the research topic the results from these studies have a strong impact not only in basic research but also in several fields of application ranging from pharmaceutical applications to the use of biomolecules as, for example, cryoprotectants.

Almost all contributions to the two volumes highlight the fact that ligand–protein complexes cannot be treated as static ensembles. On both sides, the ligand and the receptor side, dynamic processes contribute to the molecular recognition. In this sense it is hoped that these two volumes of *Bioactive Confor-*

mation will sensitize us for the need to invent and develop more experimental techniques to study the dynamic aspects of bioactive conformations.

Lübeck, December 2007 Thomas Peters

Contents

Contents of Volume 272

Bioactive Conformation I

Volume Editor: Peters, T.
ISBN: 978-3-540-49077-7

Top Curr Chem (2008) 273: 1–14
DOI 10.1007/128_2007_16

Published online: 23 April 2008

NMR-Based Strategies to Elucidate Bioactive Conformations of Weakly Binding Ligands

Marcel J. J. Blommers (✉) · Andre Strauss · Martin Geiser · Paul Ramage · Helmut Sparrer · Wolfgang Jahnke

Novartis Institutes for BioMedical Research, Discovery Technologies, 4002 Basel, Switzerland
marcel_jj.blommers@novartis.com

Abstract Key processes in molecular biology are regulated by interactions between biomolecules. Protein–protein and protein–ligand interactions, e.g., in signal transduction pathways, rely on the subtle interactions between atoms at the binding interface of the involved molecules. Because biomolecules often have many interacting partners, these interactions are not necessarily strong. The study of molecular recognition gives insight into the complex network of signaling in life and is the basis of structure-based drug design.

In the situation where the interaction is weak, one of the traditional methods that can be applied to obtain structural information (internuclear distances) of the bound ligand is the so-called transferred NOE (trNOE) method. Recently, it became possible to use transferred cross-correlated relaxation (trCCR) to directly measure dihedral angles. The combined use of these two techniques significantly improves the precision of the structure determination of ligands weakly bound to macromolecules.

The application of these techniques will be discussed in detail for a peptide derived from IKKβ bound to the protein NEMO that plays an important role in the NFκB pathway.

Keywords Cross-correlated relaxation · IKKβ · NEMO · NMR · trCCR · trNOE

Abbreviations

CCR Cross-correlated relaxation
CSA Chemical shift anisotropy

IKK	IκB kinase
NBD	NEMO binding domain
NEMO	NFκB essential modulator
NOE	Nuclear Overhauser effect
RDC	Residual dipolar coupling
SAR	Structure–activity relationship
trCCR	Transferred CCR
trNOE	Transferred NOE

1 Transferred Cross-Correlated Relaxation

1.1 Theory

Since the discovery of the nuclear Overhauser effect (NOE) [1, 2] and scalar coupling constants [3, 4] decades ago, NMR-derived structure calculations of biomolecules largely depended on measurement of these two parameters [5]. Presently, new NMR observables, like residual dipolar couplings (RDC) [6] and cross-correlated relaxation (CCR) rates [7] have become popular in conjunction with structure determination. Application of RDC provides a way to obtain long-range restraints and therefore has an impact on the resolution of biomolecular structures studied by solution-state NMR. Measurement of CCR-rates is also broadly applicable for isotopically labelled molecules and gives new direction to the design of experiments to determine the structure of biomolecules by NMR spectroscopy. The information extracted from CCR rates not only improves the resolution of NMR structures, but is also indispensable when applied to weakly binding ligands [8, 9] that are measured in the presence of the receptor using an excess of ligand. In these cases, J-coupling constants cannot be used to derive torsion angles. Alternatively, CCR-rates and RDCs provide new ways to obtain information on molecular dynamics. Herein, we will focus on CCR (for the application of transferred RDCs see the "Bioactive Conformation I" contribution by B.W. Koenig [54]).

Measurement of cross-correlated relaxation has been described for homonuclear cases [10, 11], and is widely used in solid-state NMR [12–14]. It is the availability of isotopically labelled biomolecules and its application to solution-state NMR that makes the method so interesting. The first application of CCR in solution-state NMR with a $^{15}N,^{13}C$ labelled protein, was the determination of the torsion angle ψ in the small protein rhodniin [7]. This torsion angle is difficult to obtain by traditional methods.

CCR can easily be explained in a simplified form: all coherences between nuclear spins, that finally give rise to NMR signals, relax (decay) with a certain rate, and eventually disappear. In dipolar relaxation, the relaxation of a spin is mediated by the fluctuating electromagnetic field caused by adjacent

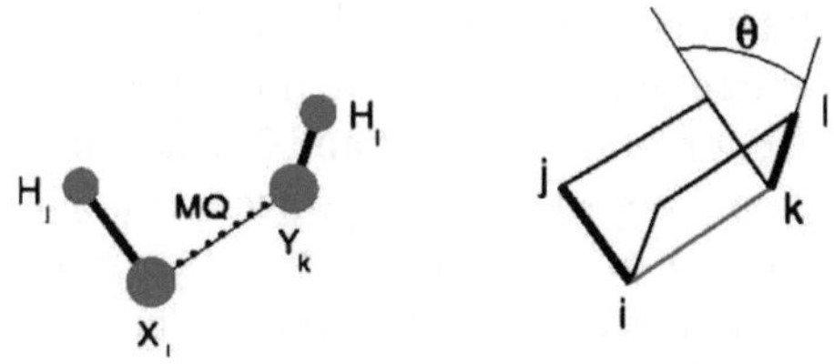

Fig. 1 Cross-correlated relaxation: Interaction between magnetic moments are dependent on the angle between them. This interaction is only present for large proteins or protein-ligand complexes

spins. Cross-correlated dipolar relaxation indicates that the dipolar coupled spin pair is not isolated, but experiences fluctuating electromagnetic fields from other spin pairs, which influence its relaxation rate. The dipolar interaction, CCR-rate, Γ_{ikjl} between these magnetic moments depends on the angle between them:

$$\Gamma_{\text{dipole–dipole}} = \frac{2}{5}\frac{\gamma_i\gamma_j}{r_{ij}^3}\frac{\gamma_k\gamma_l}{r_{kl}^3}\hbar^2\frac{1}{2}\left(3\cos^2\theta - 1\right)\tau_c\,, \tag{1}$$

where θ is the angle between the internuclear vectors i–j and k–l (Fig. 1), r denotes bond distances which are well-known, and τ_c denotes the rotational correlation time which can be measured independently.

All other factors are constants. The angle θ is the projection angle between bond vectors and is the sole unknown in Eq. 1, and can be readily and precisely determined by measurement of the cross-correlated dipolar relaxation rate. It should be emphasized that θ is measured directly, without the need of experimental calibration as in the Karplus curve for J-coupling constants for example.

Another CCR mechanism is the interaction of the magnetic dipole with the chemical shift anisotropy (CSA) tensor, e.g., the interaction with the carbonyl CSA-tensor in proteins. The dipole–CSA CCR-rate is also dependent on the projection angle θ between the magnetic dipole and CSA tensors:

$$\Gamma_{\text{dipole–CSA}} = \frac{4}{15}\varpi_j\frac{\gamma_k\gamma_l}{r_{kl}^3}\hbar\vec{\sigma}\frac{1}{2}\left(3\cos^2\theta - 1\right)\tau_c\,, \tag{2}$$

where $\vec{\sigma}$ denotes the difference $\sigma_{||} - \sigma_{\perp}$.

Cross-correlated dipolar relaxation can be measured between a variety of nuclei. The measurement requires two central nuclear spins, each of which is directly attached to a remote nuclear spin (Fig. 1). The central spin and its attached remote spin must be connected via a large scalar coupling, and the remote spin must be the primary source of dipolar relaxation for the central spin. The two central spins do not need to be scalarly coupled, although the necessity to create multiple quantum coherence between them requires them to be close together in a scalar or dipolar coupled network. In practice, the central spins will be heteroatoms (e.g. ^{13}C or ^{15}N in isotopically enriched biomolecules), and the remote spins will be their directly attached protons.

More careful inspection of Equations 1 and 2 results in the observation that these CCR-rates not only depend on the projection angle θ, but also on the rotational correlation time τ_c. This makes the determination of these rates particularly suitable for the study of torsion angles in weakly binding ligands. J-couplings cannot be used because they are equally weighted for bound and free conformations, and only a small fraction of the population ligands is bound. Therefore, a different source of information about torsion angles is required. In contrast, the CCR-rate becomes negligibly small for the unbound state, so that the CCR-rate measured for excess ligand in fast exchange with its bound state, represents essentially the bound state, since the CCR is transferred and measured on the resonances of the free ligand. This experiment therefore has been called transferred CCR (trCCR), in analogy with the transferred NOE (trNOE) experiment. We have recently reviewed these techniques for weakly binding ligands [22], i.e., transferred NOE, transferred CCR as well as transferred RDC and this work is an extension of our previous review.

1.2 Applications

The trCCR experiment has been applied to study the bound conformation of a nucleotide analog bound to Elongation Factor Tu [8, 23]. Measurements of the CCR-rates $\Gamma_{H1'C1'C2'H2'}$ and $\Gamma_{H3'C3'C4'H4'}$ of the ^{13}C-labelled nucleotide in the presence of its receptor, resulted in an unambiguous determination of the sugar conformation of the nucleotide.

Another application of trCCR is in the study of the bound conformation of the natural compound epothilone A bound to tubulin. Epothilone [24] received interest in pharmaceutical research, because of its potency as an anti-cancer drug. The research efforts include the synthesis of more than 300 epothilone derivates with the aim to find the optimal molecule for clinical application [25, 26]. It is, therefore, self-evident that there was high interest to obtain a structural model of the bioactive conformation in order to guide the lead optimization process. ^{13}C-labelled epothilone A was obtained by growing the myxobacterium *Sorangium cellulosum* on a medium containing ^{13}C-glucose. In total seven torsion angles of the macrolide polyketide have been determined in order to unambiguously derive the tubulin-bound conformation [27, 28]. Three types of CCR-rates have been measured: Γ_{CHCH} dipole–dipole CCR, Γ_{CHCH3} [29] dipole–dipole CCR and Γ_{CHCO} dipole–CSA CCR. The high-resolution structure that was derived from these angles complemented with transferred NOE-derived distance restraints, provided the first experimentally obtained bioactive conformation of the natural compound. Interestingly, the molecule experiences a conformational change upon binding tubulin.

Measurements of the trCCR-rate have also been applied to small peptidic ligands. In a proof-of-concept study we have studied the STAT-6 bound conformation of an IL-4 receptor-derived peptide [9]. For this study a chemically synthesized ^{15}N,^{13}C-labelled tetrapeptide was used in conjunction with Γ_{NHCH} dipole–dipole CCR and $\Gamma_{\mathrm{NHC'}}$ dipole–CSA CCR experiments.

The conformations of bound peptides were also studied using trCCR by Shimada and co-workers [30]. In this case peptides were labelled in *E. coli* as fusion proteins with the M13 coat protein, and were expressed in ^{15}N, ^{13}C labelled media.

2
Strategy for the Structure Determination of a Weakly Bound Peptide

Recently, we studied by trNOE and trCCR the interaction of a peptide IKK$\beta^{735-745}$ derived from IKKβ which is reported to interact with NEMO by trNOE and trCCR. The experimental data as well as the strategy to obtain the backbone torsion angles of this peptide is presented here in detail.

2.1
Biological Context

The transcription factor NF-κB plays a central role in inflammation, upregulating at least 70 genes responsible for anti-apoptotic effects in lymphocytes. It plays an important role also for cell signaling mediated by cytokines and chemokines, as well as for cell migration using adhesion molecules. This prominent role has made NF-κB an important target in inflammation and cancer [31]. The activity of the IκB kinases α and β (IKKα and IKKβ) is controlled by a scaffold protein called NEMO (NF-κB essential modulator). While the kinase activity of the IκB kinase complex (IKK) is essential to activate this pathway and has been a target in the pharmaceutical industry for a long time, a new site of interference has been described [32]. It has been shown that a short hydrophobic peptide motif present on both IκB kinases interacts with NEMO and that this interaction is essential for activation of the IKK complex. The core binding motif of the IKKβ-derived peptide or NEMO-binding domain peptide (NBD peptide, IKK$\beta^{737-742}$) consists of the amino acid sequence LDWSWL. This stretch of amino acids has been extensively characterized by point mutations and the importance of every amino acid has been addressed by alanine scanning [33]. Both tryptophan residues as well as the aspartate residue are essential. Single mutation of leucine is tolerated but double mutation is not [33]. Another class of peptides was found to interfere with the oligomerization of NEMO [34]. The pharmacological consequences of inhibiting the NEMO-IKK interaction have been tested by fusing a specialized domain originating from the *Antennapedia* homeodomain to the NBD peptide TALDWSW. This fusion peptide is able to penetrate cell membranes and enter the cytoplasm of cell lines [35]. Inhibition of NFκB activation was observed in dendritic cells after treatment with the NBD peptide [36]. Also experiments in mice have been performed with the fusion construct. Strong anti-inflammatory effects on osteoclastogenesis have been observed [37, 38], as expected for an agent inhibiting the NF-κB pathway.

2.2 NMR Studies

These in vitro and in vivo experiments clearly indicate that the interaction of the NBD peptide with NEMO has an inhibitory effect on the activation of NFκB. It would, therefore, be of interest to elucidate its NEMO-bound conformation. In a feasibility study, the binding of the peptide IKK$\beta^{735-745}$ to NEMO was measured by 1-dimensional relaxation experiments. For the wild-type sequence, binding to NEMO was observed by line broadening, enhanced T1ρ-relaxation and negative NOEs observed in waterLOGSY experiments [39], whereas for a mutant in which one of the critical amino acids was mutated, no binding was observed. It was concluded that the NBD peptide binds to NEMO in a specific manner as suggested by the published data. The observations prompted us to study the NBD peptide by the trNOE method [19–22]. The peptide was measured in 10-fold excess with respect to NEMO. Although good quality NOEs were obtained, preliminary distance-geometry calculations indicated that the structure was only poorly defined by distance restraints alone (Fig. 7A). It was, therefore, decided to label the peptide uniformly with ^{15}N and ^{13}C [40]. For these experiments we cloned the DNA fragment encoding the peptide IKK$\beta^{735-745}$ at the 3′-terminal end of the gene encoding the maltose binding protein (MBP) in the vector pXI439. In this vector the peptide and the MBP are separated by a PreScission™ protease cleavage site (Amersham Pharmacia Biotech). After expression of the fusion protein in *E. coli* BL21(DE3), by growing in minimal medium supplemented with ^{15}N ammonium chloride and ^{13}C glucose, the peptide was cleaved from MBP by digestion with the PreScission™ protease. With this protocol about 20 mg ^{15}N, ^{13}C-labelled peptide $G^{-1}P^{0}$TALDWSWLQTE, abbreviated "IKK$\beta^{735-745''}$", was obtained in a very cost-efficient way. The labelling of the peptide was aimed to complement the transferred NOE data (inter-proton distances) with transferred CCR data (backbone torsion angles).

As described in more detail below, the peptide backbone torsion angles $\psi_i(N_iC\alpha_iC'_iN_{i+1})$ were obtained by measuring the Γ_{NHCH} dipole–dipole CCR and $\Gamma_{NHC'}$ dipole–CSA CCR, which can be obtained by HN(CO)CA-derived experiments. The torsion angles $\phi_i(C'_{i-1}N_iC\alpha_iC'_i)$ were obtained by measuring the $\Gamma_{HNN\ HNH\alpha}$ dipole–dipole CCR, which can be derived from a HSQC experiment recorded at high resolution and without decoupling of ^{15}N.

2.2.1
Measurement of the Torsion Angle ψ

After the ^{15}N,^{13}C-labelled NBD peptide was obtained, the triple-resonance experiment developed to measure $\Gamma_{NHC'}$ dipole–CSA CCR-rates [41] was performed as a 2-dimensional experiment. The spectra are obtained by applying a pulse sequence that was developed by Kay and co-workers [41] to measure $\Gamma_{NHC'}$ with increased sensitivity. The experiment is one of the sensitive triple-resonance experiments, and is therefore suitable to test the feasibility of CCR measurements and to optimize sample conditions. The pulse sequence is derived from an HNCOCA and the double and zero-quantum $^{13}C'$–$^{13}C\alpha$ coherences, operative during certain delays in the experiment, experience cross relaxation and auto relaxation. The cross peaks are doublets separated by the $^{1}J_{C\alpha H\alpha}$ coupling and are observed at the C′ frequency. High-quality spectra were obtained without overlap of resonances. The intensity of the two peaks in each doublet was of the same magnitude which indicates that no cross-correlated relaxation mechanism is operative in the free peptide. This is expected, since the free peptide is small and has a relatively small τ_c. All further experiments of the peptide in the presence of NEMO were performed using a peptide concentration of 0.8 mM. After measuring the peptide in the presence of NEMO in a 10:1 ratio, no signals were observed. Apparently, auto relaxation and cross relaxation were too strong, and there was a need to optimize the experimental conditions. As is shown in Fig. 2B, with a 100-fold excess of ligand, we were able to measure the signals of eight out of ten amino acids, but the relaxation effects were still very strong. The optimal condition is 200-fold excess of ligand. In this case all signals are observed (Fig. 2A), and the CCR-rates can be accurately measured from the intensities of the single peaks. The resonance assignment as indicated follows from triple-resonance experiments measured of the free peptide. In a theoretical study, the transferred CCR-rates have been simulated [42]. It was discussed that trCCR may be more difficult to be measured than trNOE because the auto relaxation may be stronger than the CCR. Sample optimization by changing the ligand and protein ratio in order to find the optimal effective τ_c (not too small, not too high) is therefore essential as this example illustrates. Apart from changing the protein concentration, one could also change temperature, ionic strength or solvent viscosity to modulate the effective τ_c.

The individual lines of the doublet signals relax with different rates due to the phenomenon of cross-correlated dipole–CSA relaxation. The CCR-rate can be extracted from the ratio of the intensity of the peaks: $\Gamma_{\text{dipole-CSA}} = 1/2t \ln\left(I_{\text{highfield}}/I_{\text{lowfield}}\right)$, where t is the time the selected coherences experience relaxation, and I denotes the signal intensities of the spin states.

A cross section through the doublet assigned to W^{739} is shown in Fig. 3A. The CCR is separated from auto relaxation effects, and can be directly read from the signal intensities. In the case of the NBD-peptide, it is observed that the auto relaxation is not the same for all amino acids. This indicates that some amino acids experience faster relaxation caused by contacts with the protein and restricted motion. Some amino acids, e.g., T^{735} (Fig. 2), show weaker auto relaxation. These signals belong to flexible amino acids. T^{735} is one of the non-essential amino acids close to the N-terminus of the peptide.

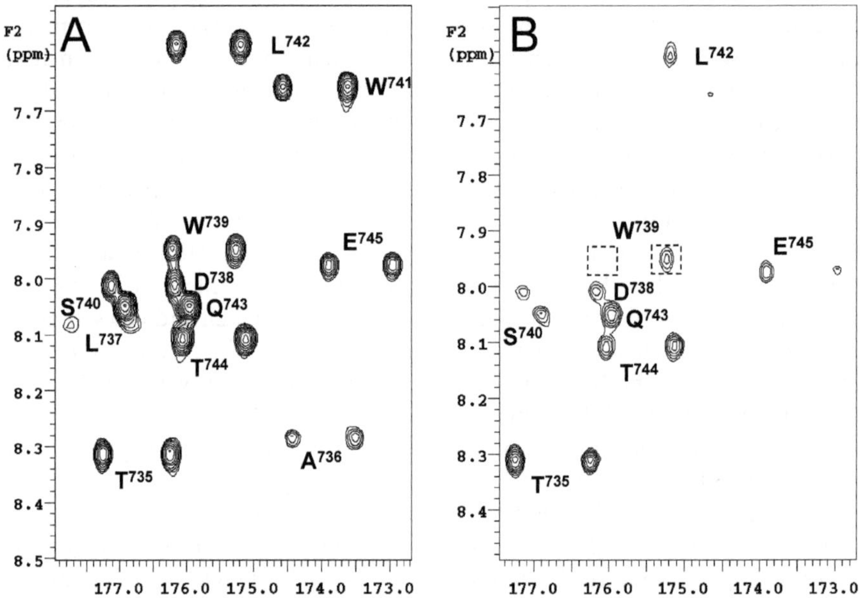

Fig. 2 HN(CO)CA-derived experiment for the measurement of $\Gamma_{NHC'}$ dipole–CSA CCR-rates. Various ratios of NBD peptide and NEMO were used. **A** 200-fold excess of ligand: optimal conditions. **B** 100-fold excess: auto relaxation and cross relaxation are too fast. The resonance assignment of the peptide is indicated

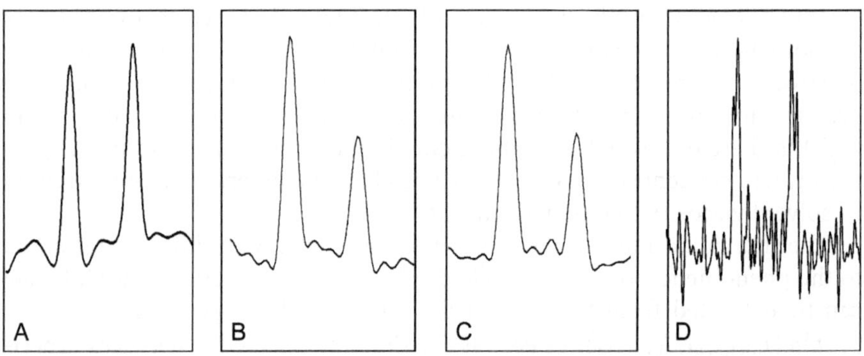

Fig. 3 Cross sections through the peaks corresponding to W^{739} in Fig. 2A and Fig. 4B,C respectively and T^{744} in Fig. 6C. CCR-rates can be deduced from the intensities of the peaks (see text). Intensities of the single lines in the absence of CCR (unbound peptide) are equal, because of the short rotational correlation time of the peptide

The same sample was subsequently used to measure Γ_{NHCH} dipole–dipole CCR. In this case the pulse sequence proposed by Yang and Kay [43] was applied. The experiment is also based on an NH(CO)CA experiment. The zero and double quantum coherences result in two 2-dimensional (2D) datasets and the 2D spectra obtained (black and red cross peaks in Fig. 4) result after pairwise adding and subtracting the measured 2D datasets. The signals are detected at the $^{13}\mathrm{C}\alpha$ frequency and split by the $^{1}\mathrm{J}_{\mathrm{C}\alpha\mathrm{H}\alpha}$ coupling. The black and red cross peaks are shifted by $^{1}\mathrm{J}_{\mathrm{NHH}\alpha}$. Also in this case, the CCR-rate can directly be obtained from the intensities of the individual peaks: $\Gamma_{\text{dipole–dipole}} = 1/4t \ln\left(I^{\text{red}}_{\text{highfield}}/I^{\text{red}}_{\text{lowfield}} \cdot I^{\text{black}}_{\text{lowfield}}/I^{\text{black}}_{\text{highfield}}\right)$.

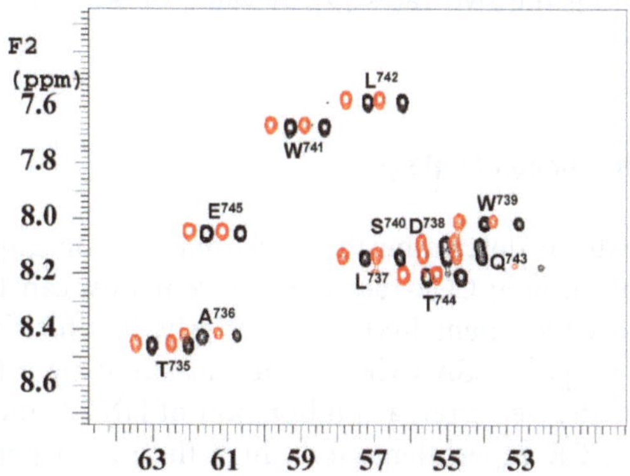

Fig. 4 HN(CO)CA-derived experiment to measure CH-NH dipole–dipole CCR-rates recorded for the uniformly ^{15}N, ^{13}C-labelled NBD peptide bound to NEMO. A 200-fold excess of peptide was chosen. The resonance assignment is given

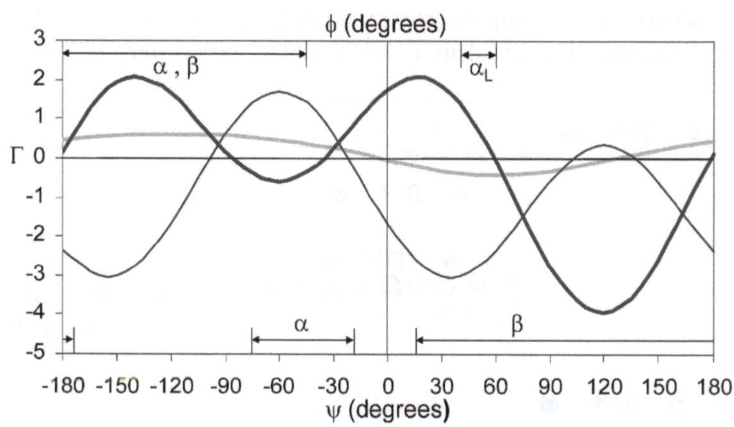

Fig. 5 Dependence of the CCR-rates on the peptide backbone torsion angles Ψ(N-Cα-C′-N) and ϕ(C′-N-Cα-C′). The *thin curve* corresponds to the dipole–CSA CCR-rate ($\Gamma_{\mathrm{NHC'}}$), the *thick curve* to the dipole–dipole CCR-rate (Γ_{NHCH}) [9] and the *grey curve* to the dipole–dipole CSA-rate ($\Gamma_{\mathrm{HNNHNH}\alpha}$) [47]. The allowed regions for Ψ and ϕ are indicated; α indicates α-helix, α_{L} is left-handed helix, β is β-sheet

In Fig. 3B and 3C cross sections of the doublet in the black and red spectrum corresponding to W^{739} are shown.

The peptide backbone torsion angle ψ was determined by combining the two relaxation rates obtained by the above-described experimental data. The dependence of the CCR-rates on the ψ-values is given in Fig. 5. Analysis of the individual lines corresponding to W^{739} (cross sections in Fig. 3) gives rates of 0.21 and −0.65 Hz for the dipole–dipole and dipole–CSA CCR-rates, respectively. The value of the torsion angle ψ is obtained by calculating $\Gamma_{\text{dipole–dipole}}/\Gamma_{\text{dipole–CSA}}$ for the allowed region for ψ where $\Gamma_{\text{dipole–dipole}}$ is positive and $\Gamma_{\text{dipole–CSA}}$ is negative. The effective correlation time is not required by combining the two rates [9]. A value of $\Psi = 35°$ is found for the residue W^{739}.

2.2.2
Measurement of the Torsion Angle ϕ

A suitable CCR-rate to determine the backbone torsion angle ϕ by CCR is the $\Gamma_{\text{NHCH}\alpha}$ dipole–dipole CCR-rate that conveniently can be measured by an HNCA-derived experiment [44]. Alternatively, like for the torsion angle ψ, the $\Gamma_{\text{CH}\alpha\text{C}'(i-1)}$ dipole–CSA CCR can be measured by a triple-resonance experiment that is derived from a combination of HNCA and HNCO experiments [45]. Also, CCR experiments for which the rate depends on ψ and ϕ simultaneously are available, e.g., the $\Gamma_{\text{CH}\alpha(i)\text{CH}\alpha(i-1)}$ dipole–dipole CCR experiment can be used [46]. Unfortunately for the peptide under investigation, we were not able to successfully record any of these spectra, possibly due to the relatively strong auto relaxation.

A less obvious dipole–dipole interaction is the $\Gamma_{\text{HNNHNH}\alpha}$ CCR. This CCR-rate is conveniently measured in a $^1H^{15}N$ HSQC experiment [47] for which

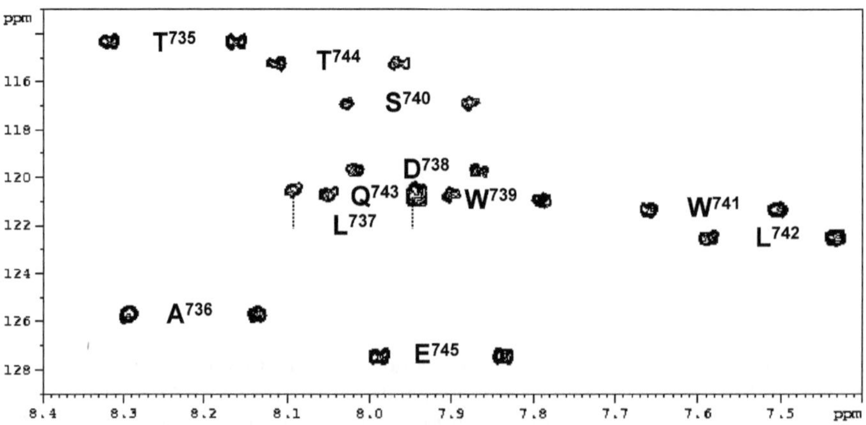

Fig. 6 HSQC experiment of the NBD peptide bound to NEMO for the measurement of HNN-HNHα dipole–dipole CCR. The resonance assignment is indicated

^{15}N is not decoupled during acquisition. A prerequisite is that the experiment is recorded at high resolution in the ^{1}H dimension and that the lines are not too broad. The CCR-rate can be derived from the intensity of the four single lines by $\Gamma_{\text{dipole-dipole}} = 1/4\left(\Sigma I_{\text{innerlines}} - \Sigma I_{\text{outerlines}}\right)$.

A positive CCR-rate corresponds to a negative value of ϕ (e.g., α-helix, turn or β-sheet), a negative CCR-rate corresponds to a positive value of ϕ (e.g. turn or L-helix) (Fig. 5). Although an accurate value of the torsion angle could not be obtained, at least these loose restraints can be used for structure determination if required. The method is easily applied and less sensitive to auto relaxation. An example is the HSQC spectrum recorded for the NBD peptide in the presence of NEMO displayed in Fig. 6. A cross section corresponding to T^{744} is shown in Fig. 3D. For this case, the inner lines are stronger than the outer lines, indicating a negative value for the torsion angle ϕ.

2.3 Structure Calculations

Using 18 trNOE-derived distance restraints and 13 trCCR-derived backbone torsion angle restraints, structure calculations using distance geometry and simulated annealing [48] resulted in a well-defined structure of the IKKβ-derived peptide bound to NEMO. The backbone structure is displayed in Fig. 7B and is compared with the result of the calculation carried out using the trNOE-derived distance restraints alone. It is obvious from Fig. 7 that only the combination of the trNOE- and trCCR-derived restraints results in the structure elucidation of the bound conformation of this peptide.

3 Conclusion and Outlook

In cases where the interactions are weak, and the exchange between the free and bound state is fast, trNOE and trCCR are important NMR techniques to study the bound state, particularly because the measured NOE or CCR-rate is almost exclusively built-up when the ligand is bound. The bioactive conformation of weakly binding ligands can provide information which can be used to understand protein–ligand interactions, and can provide information to design mimics of the ligand which may serve as a drug. In the latter case, one could design molecules using the conformation of the bound ligand together with SAR information of analog ligands [49], or alternatively, one can search in 3D-databases for molecules that would have the same 3D pharmacophore [50].

A more complete picture of the protein–ligand interaction is given by the docking mode of the bioactive conformation in the binding pocket of the receptor protein. In case the apo protein structure is known from NMR or

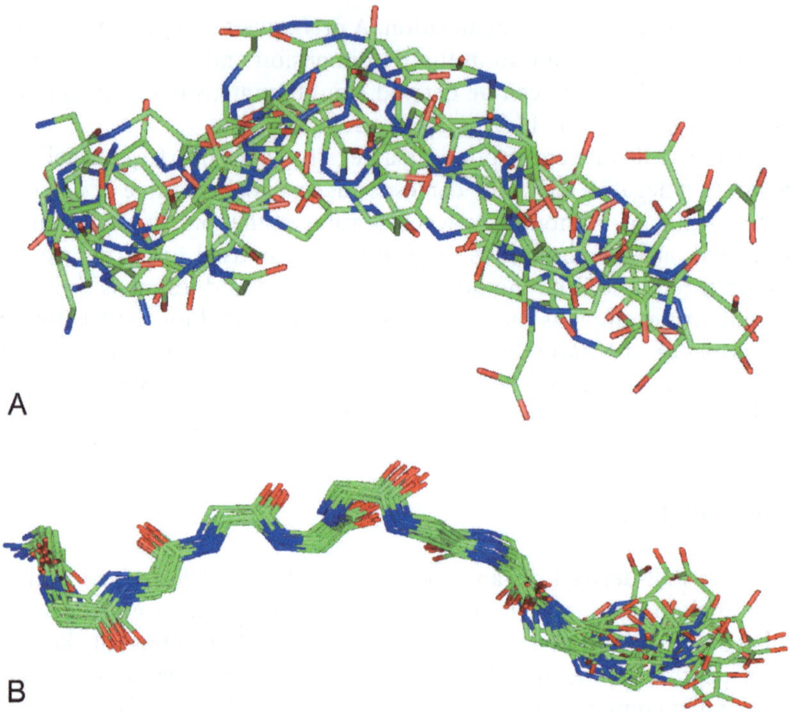

Fig. 7 Solution of structure calculations of the NBD-peptide bound to NEMO. **A** Only trNOE-derived distance restraints were used in the calculation. The structure is not defined by the distance restraints alone. **B** trNOE-derived distance restraints and trCCR-derived torsion angle restraints were combined. These restraints complement each other and are sufficient to define the NEMO-bound conformation of the IKKβ-derived peptide. In both cases ten solutions were superimposed

X-ray crystallography, docking modes can be validated by various NMR techniques: NOEs may be observed between the ligand and the receptor protein by heteronuclear-filtered NOE spectroscopy [51], chemical shift changes of protein resonances upon binding can be analyzed by simulation of shifts caused by ring currents and electrostatic effects [52], and saturation transfer difference measurements indicate which part of the ligand is in direct contact with the protein [52].

References

1. Macura S, Ernst RR (1980) Mol Phys 41:95
2. Neuhaus D, Williamson M (1989) The Nuclear Overhauser Effect in Structural and Conformational Analysis. VCH Publishers, New York
3. Karplus M (1959) J Chem Phys 30:11
4. Bystrov VF (1976) Prog NMR Spect 10:41

5. Wüthrich K (1986) NMR of Proteins and Nucleic Acids. Wiley, New York
6. Tjandra N, Omichinski JG, Gronenborn AM, Clore GM, Bax A (1997) Nat Struct Biol 4:732
7. Reif B, Hennig M, Griesinger C (1997) Science 276:1230
8. Carlomagno T, Felli IC, Czech M, Fischer R, Sprinzi M, Griesinger C (1999) J Am Chem Soc 121:1945
9. Blommers MJJ, Stark W, Jones CE, Head D, Owen CE, Jahnke W (1999) J Am Chem Soc 121:1949
10. Dalvit C, Bodenhausen G (1988) J Am Chem Soc 110:7924
11. Dalvit C, Bodenhausen G (1990) Adv Magn Res 14:1
12. Feng X, Lee YK, Sandström S, Eden M, Maisel H, Sebald A, Levitt MH (1996) Chem Phys Lett 257:314
13. Feng X, Verdegem PJE, Lee YK, Sandström D, Eden M, Bovee-Geurts P, de Grip WJ, Lugtenburg J, de Groot HJM, Levitt MH (1997) J Am Chem Soc 119:6853
14. Schmidt-Rohr K (1996) J Am Chem Soc 118:7601
15. Cavanagh J, Fairbrother WJ, Palmer AG III, Skelton NJ (1996) Protein NMR Spectroscopy. Academic Press, San Diego, p 265
16. Reif B, Diener A, Henning M, Maurer M, Griesinger C (2000) J Magn Res 143:45
17. Tessari M, Vis H, Boelens R, Kaptein R, Vuister G (1997) J Am Chem Soc 119:8985
18. Yang D, Konrat R, Kay LE (1997) J Am Chem Soc 119:11938
19. Ni F (1994) Progr NMR Spectros 26:517
20. Lian LY, Barsukov IL, Sutcliffe MJ, Sze KH, Roberts GCK (1994) Methods Enzymol 239:657
21. Campbell AP, Sykes BD (1993) Annu Rev Biophys Biomol Struct 22:99
22. Blommers MJJ, Rüdisser S (2002) BioNMR in drug research. In: Zerbe E (ed) Methods and Principles in Medicinal Chemistry. Wiley-VCH, Weinheim, Germany, p 355
23. Felli IC, Richter C, Griesinger C, Schwalbe H (1999) J Am Chem Soc 121:1956
24. Höfle G, Bedorf N, Gerth K, Reichenbach H (1993) Chem Abstr 120:52841
25. Nicolau KC, Roschangar F, Vourlominis D (1998) Angew Chem Int Ed 37:2014
26. Wartmann M, Altmann K-H (2002) Curr Med Chem Anti-Cancer Agents 2:123
27. Carlomagno T, Blommers MJJ, Meiler J, Jahnke W, Schupp T, Petersen F, Schinzer D, Altmann K-H, Griesinger C (2003) Angew Chem Int Ed 42:2511
28. Carlomagno T (2005) Ligand-target interactions: what can we learn from NMR? Annu Rev Biophys Biomol Struct 34:245
29. Carlomagno T, Sanchez VM, Blommers MJJ, Griesinger C (2003) Angew Chem 42:2515
30. Mizukoshi Y, Takahashi H, Shimada I (2006) J Biomol NMR 34:23
31. Karin M, Yamamoto Y, Wang QM (2004) Nat Rev Drug Discov 3:17
32. May MJ, D'Acquisto F, Madge LA, Glockner J, Pober JS, Ghosh S (2000) Science 289:1550
33. May MJ, Marienfeld RB, Ghosh S (2000) J Biol Chem 27:45992
34. Agou F, Courtois G, Chiaravalli J, Baleux F, Coic YM, Traincard F, Israel A, Veron M (2004) J Biol Chem 279:54248
35. Choi M, Rolle S, Wellner M, Cardoso MC, Scheidereit C, Luft FC, Kettritz R (2003) Blood 102:2259
36. Tas SW, de Jong EC, Hajji N, May MJ, Ghosh S, Vervoordeldonk MJ, Tak PP (2005) Eur J Immunol 35:1164
37. Dai S, Hirayama T, Abbas S, Abu-Amer Y (2004) J Biol Chem 279:37219
38. di Meglio P, Ianaro A, Ghosh S (2005) Arthritis Rheum 52:951
39. Jahnke W, Widmer H (2004) Cell Mol Life Sci 61:580

40. LeMaster DM (1994) Prog NMR Spectrosc 26:371
41. Yang D, Gardner KH, Kay LE (1998) J Biomol NMR 11:213
42. Ravindranathan S, Mallet J-M, Sinay P, Bodenhausen G (2003) J Magn Res 163:199207
43. Yang D, Kay LE (1998) J Am Chem Soc 120:9880
44. Pelupessy P, Chiarparin E, Ghose R, Bodenhausen G (1999) J Biomol NMR 14:277
45. Kloiber K, Konrad R (2000) J Biomol NMR 17:265
46. Chiarparin E, Pelupessy P, Ghose R, Bodenhausen G (2000) J Am Chem Soc 122:1758
47. Crowley P, Ubbink M, Otting G (2000) J Am Chem Soc 122:2968
48. Havel TF (1991) Prog Biophys Mol Biol 56:43
49. Bohm H-J, Schneider G (2003) Protein-Ligand Interactions From Molecular Recognition to Drug Design. Methods Princ Med Chem, vol 19. Wiley-VCH, Weinheim, Germany
50. Alvarez J, Shoichet B (2005) Virtual Screening in Drug Discovery. CRC-Press, Boca Raton, FL
51. Otting G, Wüthrich K (1990) Quart Rev Biophys 23:39
52. McCoy MA, Wyss DF (2000) J Biomol NMR 18:189
53. Mayer M, Meyer B (2001) J Am Chem Soc 123:6108
54. Koenig BW (2007) Top Curr Chem 272:187

Top Curr Chem (2008) 273: 15–54
DOI 10.1007/128_2007_144

Published online: 30 August 2007

Quantitative Analysis of STD-NMR Spectra of Reversibly Forming Ligand–Receptor Complexes

N. Rama Krishna (✉) · V. Jayalakshmi

Department of Biochemistry and Molecular Genetics, NMR Core Facility, Comprehensive Cancer Center, University of Alabama at Birmingham, Birmingham, AL 35294-2041, USA
nrkrishna@bmg.bhs.uab.edu

Abstract We describe our work on the quantitative analysis of STD-NMR spectra of reversibly forming ligand–receptor complexes. This analysis is based on the theory of complete relaxation and conformational exchange matrix analysis of saturation transfer (CORCEMA-ST) effects. As part of this work, we have developed two separate versions of the CORCEMA-ST program. The first version predicts the expected STD intensities for a given model of a ligand–protein complex, and compares them quantitatively with the experimental data. This version is very useful for rapidly determining if a model for a given ligand–protein complex is compatible with the STD-NMR data obtained in solution. It is also useful in determining the optimal experimental conditions for undertaking the STD-NMR measurements on a given complex by computer simulations. In the second version of the CORCEMA-ST program, we have implemented a torsion angle refinement feature for the bound ligand within the protein binding pocket. In this approach, the global minimum for the bound ligand conformation is obtained by a hybrid structure refinement protocol involving CORCEMA-ST calculation of intensities and simulated annealing refinement of torsion angles of the bound ligand using STD-NMR intensities as experimental constraints to minimize a pseudo-energy function. This procedure is useful in refining and improving the initial models based on crystallography, computer docking, or other procedures to generate models for the bound ligand within the protein binding pocket compatible with solution STD-NMR data. In this chapter we describe the properties of the STD-NMR spectra, including the dependence of the intensities on various parameters. We also describe the results of the CORCEMA-ST analyses of experimental STD-NMR data on some ligand–protein complexes to illustrate the quantitative analysis of the data using this method. This CORCEMA-ST program is likely to be useful in structure-based drug design efforts.

Keywords STD-NMR · CORCEMA-ST · SICO · Bound ligand conformation · Structure-based drug design

1 Introduction

Biological molecules generally exert their activities through the formation of complexes with other macromolecules or with low molecular weight ligands. The formation of such complexes and the associated interactions are important mechanisms by which cellular functions are regulated. They are involved in many cellular processes such as the regulation of hormones, gene expression, signaling, modulation of ion channels, immune response, and enzyme catalysis, to name a few. Some of these macromolecules can often serve as suitable targets for a strategic manipulation of their activities through the design of low molecular weight pharmaceutical drugs. The ultimate design of such drugs is often facilitated by a screening of suitably designed large compound libraries to identify promising lead compounds that are specific to the target protein of interest, followed by a structure-based drug design from a knowledge of the bound structure of the lead compound within the active site of the target protein. In this regard, NMR spectroscopy is now playing an increasingly important role.

A large number of NMR-based screening methods have been developed to identify bioactive ligands. They include methods based on, e.g., trNOESY [1],

NOE pumping [2] and reverse NOE-pumping [3], STD-NMR [4–6], WaterLOGSY [7, 8], relaxation filtering [9, 10], diffusion editing [10, 11], SHAPES screening [12], flow-injection [13], chemical shift perturbations [14–16], and methods based on target protein signals [17], to name a few (also see related chapters in the preceding volume by Steven LaPlante, Bernd Reif, as well as Timo Weide, Armin Modlinger and Horst Kessler, and in this volume by Jesus Jimenez-Barbero). The advantages and limitations of each of these and other NMR-based screening methods have been reviewed [18–22]. Among these, the recently introduced STD-NMR technique is gaining popularity because of its high sensitivity in detecting binding to a protein using, if necessary, only a trace quantity of a target protein (e.g., with ligand/protein ratios in the range of ∼50 : 1 to ∼1000 : 1), and its ability in identifying the group epitopes of the bound ligand in a straightforward but qualitative manner. This method is best suited when the binding is relatively weak, with dissociation constants (K_d) in the range $\sim 10^{-3}$ M to $\sim 10^{-7}$ M. The application of this method in screening compound libraries [4–6], and in the identification of group epitopes of the bound ligands to proteins [23–26], nucleic acids [27] and to virus coat proteins [28] has been demonstrated.

STD-NMR is also used to characterize binding interactions of cell surface proteins in living cells [29, 30]. In some recent studies, the structures of a protein–ligand complex was solved by performing STD-NMR spectroscopy on a ligand complexed to a series of target samples that had been perdeuterated everywhere except for specific amino acid types [31]. In this way the amino acid composition of the ligand binding site was defined and given the three dimensional (3D) structure of the protein target, the 3D structure of the protein–ligand complex can be determined. The advantage of this method is that no protein resonance assignments are necessary a priori for solving the structure of the complexes since these are inferred from the specific proton-labeling of an otherwise perdeuterated protein. The disadvantage is the experimental difficulties involved in the preparation of series of perdeuterated target samples with specific residues in protonated form. STD-NMR was also used for the assignment of downfield proton resonances in an enzyme inhibitor complex [32].

X-ray crystallography has traditionally been the source of information in structure-based drug design efforts in the pharmaceutical industry, with solution NMR spectroscopy often playing a complementary role. However, in those instances where it is not possible to get high quality single crystals of the complexes, solution NMR is the *only* method that is available for obtaining detailed structural information. Many of the initial lead compounds to a specific target protein are likely to exhibit only weak binding, with affinities in the milli- to micromolar range. Thus, in this regard, NMR spectroscopy provides a very powerful avenue for gaining the structural information of these reversibly forming weak complexes. Here we describe the results of recent efforts in our laboratory on gaining such structural information on weak complexes.

Our laboratory has extended the complete relaxation and conformational exchange matrix (CORCEMA) theory [33–35] to STD-NMR experiments [36, 37] and has shown that the STD-NMR intensities exhibit a complex dependence on a variety of factors such as the correlation times and relaxation times, conformations of the ligand and the protein in their free and bound states, the binding constant, exchange-mediated leakage, saturation time, and the location of the saturated protein proton(s) in the ligand–protein complex with respect to the bound ligand. [36, 37] We have developed two versions of the program that implemented the CORCEMA analysis of STD data. The first version of the CORCEMA-ST (ST, saturation transfer) program [36] could be used to simply predict the STD-NMR intensities for any given model of a ligand–receptor complex. Thus, it can be readily used to verify if the experimental STD-NMR intensities are compatible with a crystallographic model or, in its absence, with a computational model generated by a docking program. We extended our theoretical work and developed the second version of the CORCEMA-ST program to demonstrate that it may be possible to determine the bound-conformation of a weakly binding ligand within the binding pocket of a target protein from a CORCEMA-based optimization of an NOE R-factor to get the best fit between experimental STD-NMR values and calculated values. In our approach, the global minimum for the bound-ligand conformation is obtained by a combined CORCEMA and simulated annealing refinement of torsion angles of the bound ligand using STD-NMR intensities as experimental constraints. Additionally, if necessary, one can optimize other parameters such as correlation times and binding constants to obtain best fits between calculated and experimental STD-NMR values.

This review is a brief summary of the theory and application of the CORCEMA-ST protocol in the quantitative analysis of STD-NMR data. The reader is referred to the various original articles for the details.

1.1
STD-NMR Experiments

In STD-NMR experiments, one generally works with a solution containing an excess of a single ligand or a library of compounds in the presence of a trace amount of a target protein (typically with ligand/receptor ratios ranging from 10 : 1 to 1000 : 1). The NMR signal(s) of the target molecule is saturated by rf irradiation without directly affecting the NMR signals from the low molecular weight ligand. Under weak binding conditions, the saturation spreads to the remaining protein protons as well as to the ligand protons in the free and bound states through a network of dipolar and conformational exchange processes. [36] The STD-NMR spectra are obtained by recording 1D-NMR spectra of the ligand with and without saturation of the protein protons, and taking a difference. In the 2D-NMR implementation, the 2D-TOCSY spectra are obtained with and without irradiation of the resonances of the macromolecule and a difference obtained. In the STD-NMR experiments only those protons within the ligand that are at the interface with the protein are generally expected to exhibit significant intensity changes and, hence, the ligand surface epitopes are easily identified.

1.2 ICS-NMR Experiments

The CORCEMA-ST theory is also applicable for intermolecular cross-saturation NMR (ICS-NMR) experiments used for mapping the protein–protein contact surface [38, 39] or protein–RNA contact surface in a complex [40], though the discussion in this chapter is devoted to STD-NMR experiments only. In the ICS experiments [38, 39], the protein of interest is uniformly labeled with ^{15}N and ^{2}H (which eliminates all aliphatic hydrogens from this protein). The [^{1}H-^{15}N] correlation NMR spectra of this labeled protein are then recorded with and without selective saturation of the aliphatic proton resonances of the unlabeled large protein within the complex. The amide protons from those amino acids within the labeled protein that are at the protein–protein interface will experience greater saturation transfer and hence their correlation peaks will exhibit significantly reduced intensity changes, thus facilitating an easy identification of residues at the interface. Deuteration is not required to map the interface in RNA–protein complexes [40]. Though these ICS-NMR experiments are traditionally performed on tight complexes, they are also applicable to weak protein–protein complexes [39].

2 CORCEMA-ST Theory

The basic theoretical formulation for the quantitative analysis of the STD-NMR data is similar to the one we developed for the analysis of transferred NOESY data, and is based on a complete relaxation and conformational exchange matrix (CORCEMA) theory [33].

Here, we present a formulation suitable for reversible binding of a ligand and a protein to form a binary complex. The two-state kinetic model involves the interacting molecular species $\boldsymbol{L}$ and $\boldsymbol{P}$ under equilibrium conditions, exchanging between their respective free states and the bound state $\boldsymbol{L'P'}$ according to Scheme 1.

We assumed that the receptor protons can be subdivided into two classes, $\boldsymbol{P1}$ (and $\boldsymbol{P1'}$) and $\boldsymbol{P2}$ (and $\boldsymbol{P2'}$), where $\boldsymbol{P2}$ and $\boldsymbol{P2'}$ stand for protons that experience rf irradiation directly (fast exchange on the chemical shift scale is assumed). The primes indicate protons within the complex. $\boldsymbol{P1}$ and $\boldsymbol{P1'}$ stand for all the remaining protons in the receptor in its free and bound states, re-

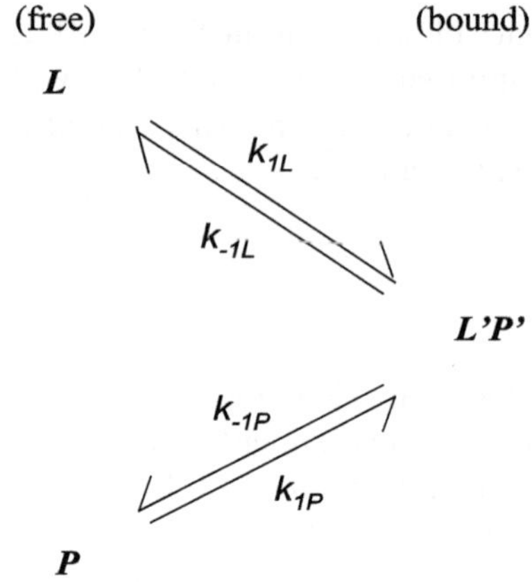

Scheme 1

spectively. Thus, in our model, the saturation originates from ***P2*** and ***P2'*** and spreads to all other protons (ligand and protein) according to Scheme 2 (DD stands for exchange through dipole–dipole relaxation).

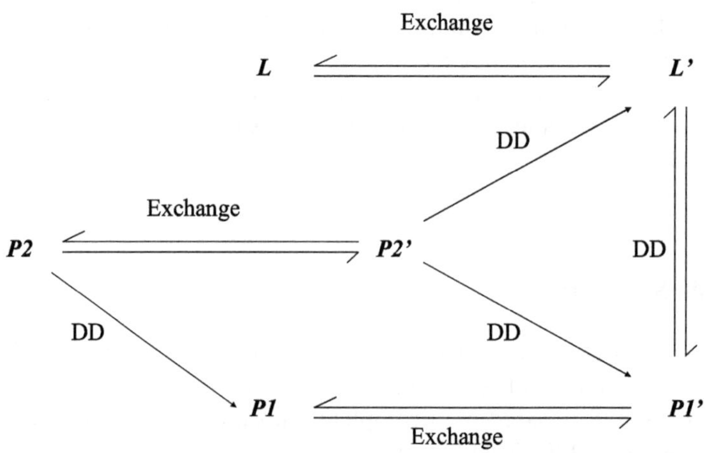

Scheme 2

Under these conditions, the expression for the observable magnetization in an intermolecular NOE (INOE) experiment in which the ***P2*** and ***P2'*** protons are *instantaneously* saturated by rf irradiation is given by [36]:

$$\mathrm{I}(t) = \mathrm{I}_0 + \left[1 - \exp\left\{-\left(\mathbf{R} + \mathbf{K}\right)t\right\}\right](\mathbf{R} + \mathbf{K})^{-1}\mathbf{Q}\,, \tag{1}$$

where t is the time period for which the protons remain saturated. This expression is a solution of a set of differential equations obtained from standard equations of motions for the magnetizations coupled by dipolar and chemical *exchange processes after setting the magnetizations of saturated receptor protons (P2 and P2′) to zero.* All quantities in bold letters (such as **I**, **R**, **K**, **Q** etc.) stand for matrices or group of protons. The generalized kinetic matrix **K**, relaxation matrix **R**, intensity matrix **I** and **Q** have been defined earlier [36]. The dynamic matrix **R** + **K** is asymmetric, and can be put in a symmetric form for convenient numerical diagonalization purposes using a symmetrization matrix **S** [33]. Thus, if Λ is the diagonalized form of the **R** + **K** matrix:

$$\Lambda = \mathbf{T}^{-1}\mathbf{S}^{-1}(\mathbf{R}+\mathbf{K})\mathbf{S}\mathbf{T}\,, \qquad (2)$$

where **T** is the matrix that diagonalizes the symmetric matrix $\mathbf{S}^{-1}$ (**R** + **K**) **S**. Thus:

$$\mathbf{I}(t) = \mathbf{I}_0 + [1 - \mathbf{S}\mathbf{T}\exp\{-\Lambda t\}\mathbf{T}^{-1}\mathbf{S}^{-1}](\mathbf{S}\mathbf{T}\Lambda^{-1}\mathbf{T}^{-1}\mathbf{S}^{-1})\mathbf{Q}\,. \qquad (3)$$

From Eqs. 1 or 3 it is easy to compute the absolute intensity $[\mathbf{I}_{0(\mathrm{k})} - \mathbf{I}(t)_{(\mathrm{k})}]$ or fractional intensity change $[\mathbf{I}_{0(\mathrm{k})} - \mathbf{I}(t)_{(\mathrm{k})}/\mathbf{I}_{0(\mathrm{k})}]$ for the STD-NMR signal for proton k. Even though the expressions above are given for the analysis of 1D STD-NMR measurements, they are easily adapted for 2D STD-NMR measurements as well.

3 Method

Figure 1 shows the CORCEMA-ST protocol that employs Eqs. 1 or 3 (a choice is provided in the protocol) to compute the absolute and the fractional intensity of 1D STD-NMR spectra. A program employing this protocol using MATLAB (The Mathworks, Natick, MA) was written for use on a personal computer. Our experience shows that MATLAB can directly handle asymmetric matrices of the type **R** + **K** without a need for symmetrization, and that the results obtained with Eqs. 1 and 3 are identical. The starting point of the program is reading the PDB coordinates for the free ligand, free protein, and the ligand–protein complex either from existing models from the protein data bank (PDB) or from computer-generated models. This program also allows the conformations of the interacting molecules to be different, if necessary, in the free and bound states. Other input parameters consist of the total number of the ligand protons (N), and the protein protons (M) near the active site to be included in the CORCEMA calculations, the rotational correlation times, the kinetic data (i.e., equilibrium constant K_{eq} (= $1/K_{\mathrm{d}}$ where K_{d} is the dissociation constant) for the complex together with estimates of the k_{on} or k_{off} rates), and an estimate for the non-specific leakage relaxation rate of the individual protons in the free and bound states [34]). Additionally, in setting up the **R** matrix, flags are set to identify situations that require a consideration of internal motions. Specifically, the methyl group protons (from Leu, Val, Thr, Ala, and Ile) and the aromatic ring protons (from Tyr and Phe

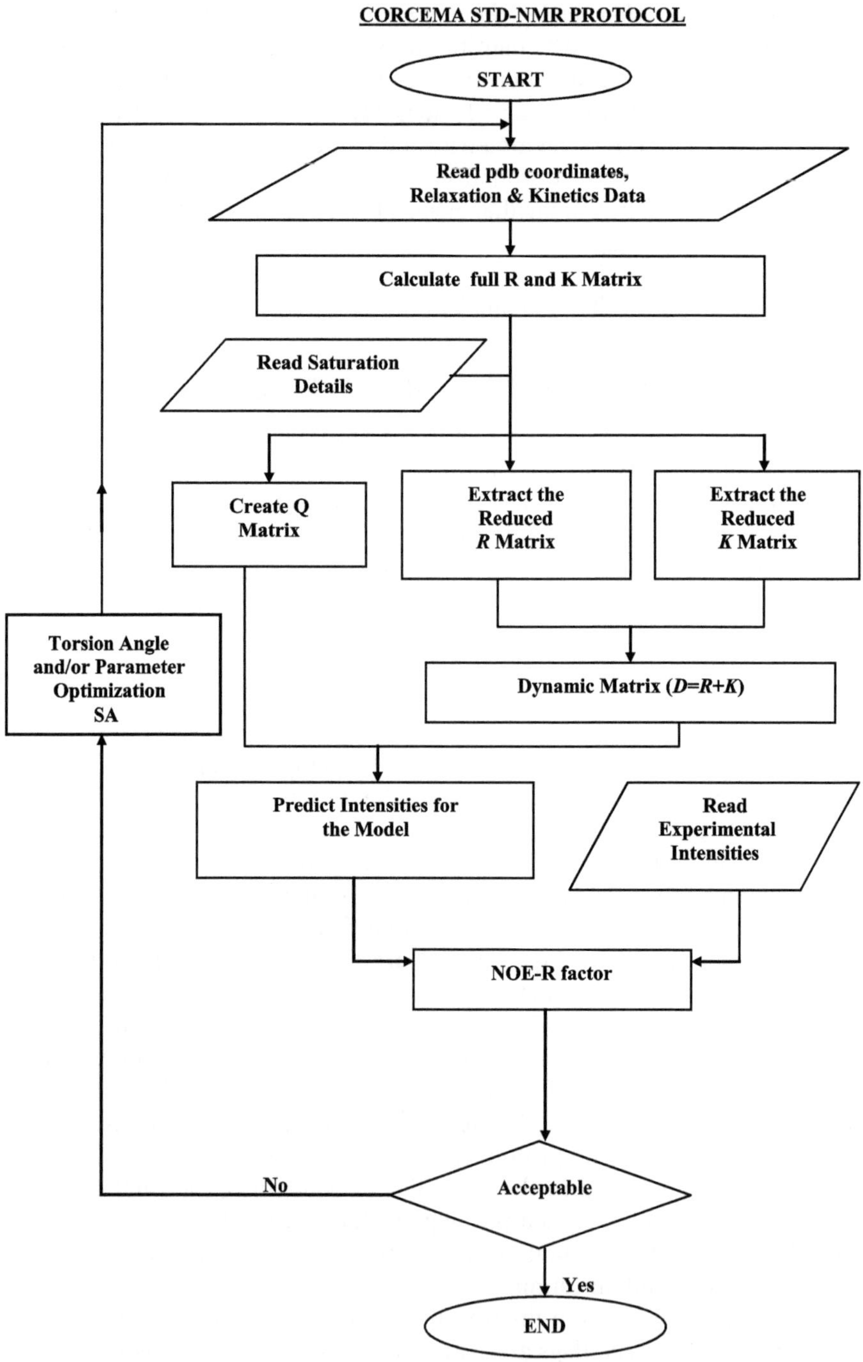

Fig. 1 CORCEMA-ST protocol for calculating STD-NMR intensities and for optimizing the bound-ligand conformation. Reprinted from [49]; © 2004, with permission from Elsevier

with twofold symmetry) are identified. For methyl groups, where the internal rotational correlation times can be considerably shorter than the overall rotational correlation time, a "model-free" calculation of spectral densities with order parameters (S^2) and internal (τ_m) and overall (τ_r) correlation times was used to calculate spectral densities involving intramethyl and methyl-to-other proton dipolar interactions, as in [34, 35, 41–43]. For intramethyl proton–proton interactions, S^2 was set to 0.25. For methyl-to-other proton interactions, the S^2 value was a read-in value, but generally kept in the range of 0.6 to 0.9 [34, 42, 44]. For Tyr and Phe, it was assumed that the ring-flip rate was much longer than the rotational correlation time; a simple $\langle 1/r^6 \rangle$ average was used for the interaction of external protons with the equivalent protons in these two aromatic residues.

The program first sets up sparse matrices for **R** and **K**. From a knowledge of the identity of the saturated receptor protons, the **Q** matrix consisting of cross-relaxation elements involving ***P2′*** (and ***P2***), and the reduced **R** and **K** matrices corresponding to ***L***, ***L′***, ***P1*** and ***P1′*** protons, are created. The dimensions of the reduced **R** and **K** matrices (Eq. 1) are $2(N + M - J)$ where J is the number of ***P2*** protons. The dimension of the Q-matrix is $2(N + M - J) \times 1$. To speed up the computations considerably on a PC, the program also gives an option of using an arbitrary cut-off distance (r_{cutoff}, usually 10 Å) in computing the **R** matrix, i.e., if the distance between any pair of intra- or intermolecular hydrogens is > 10 Å, the corresponding element in the relaxation rate matrix is set automatically to zero without further calculation. The concentrations of the species are calculated using standard equations (e.g., see equations on p 254 of [33]). The elements of $\mathbf{I}_0$ are set to be identical to the concentrations of the molecular species. In the final step, the "predicted" STD changes are calculated and compared with the experimental values using the NOE R-factor [45, 46].

3.1 Experiments with Finite Delays

Equation 1 was derived assuming that the nuclear spin magnetizations are at thermal equilibrium values prior to the start of the presaturation. In practice, due to time constraints on the instrument, this condition may not usually be realized and the nuclear spin magnetization can generally be in a quasi-equilibrium state prior to presaturation. If $(t_d + t)$ is the delay between two consecutive 90° observe pulses, where t is the presaturation period and t_d is the time delay before presaturation (this includes the data acquisition time for the previous pulse), then the appropriate expressions for STD and for control NMR spectra are given by:

$$\mathbf{I}(t) = \mathbf{I}_0 + [1 - \exp\{-(\mathbf{R}+\mathbf{K})t\}](\mathbf{R}+\mathbf{K})^{-1}\mathbf{Q} + \exp\{-(\mathbf{R}+\mathbf{K})t\}[\mathbf{I}(0)_r - \mathbf{I}_0]\,, \tag{4}$$

where

$$I(0)_r = \{[1 - \exp\{-(R_F + K_F)t_d\}]I_{0F}\}_r ,$$

and

$$I^{control} = \{[1 - \exp\{-(R_F + K_F)(t_d + t)\}]I_{0F}\}_r . \quad (5)$$

In the above expressions, the subscript F refers to the full relaxation and exchange matrices that include the ***P2*** and ***P2′*** protons since their magnetizations do not experience rf saturation during t_d in the on-resonance saturation experiment and during $t_d + t$ in the off-resonance irradiation experiment (control spectrum). Hence, experience coupled recovery with the rest of the protons during these periods. The subscript r refers to the reduced matrix containing elements for ***L***, ***L′***, ***P1***, and ***P1′*** extracted from the full matrix. We have implemented the above expressions as an option in the CORCEMA-ST program.

3.2 Non-instantaneous Saturation

Even though the assumption of *instantaneous* saturation of the irradiated proton signal is generally easily met under standard experimental conditions by adjusting the bandwidth and strength of the irradiating rf field, a provision is also provided in the protocol to treat in a phenomenological way those situations where the saturated proton(s) does not experience instantaneous saturation, but experiences saturation while undergoing Torrey oscillations [47, 48].

4 Analysis of Some Critical Factors Using Simulations of STD Data on Hypothetical Models

We carried out some simulations using two hypothetical models consisting of five ligand protons and five protein protons with the configurations shown in Fig. 2a (the symmetric model) and Fig. 2b (the asymmetric model).

In this simulation, a ligand/protein ratio of 10 : 1 was assumed. The k_{on} rate was assumed to be $10^8\ s^{-1}\ M^{-1}$. The spectrometer frequency was set at 600 MHz. A somewhat longer correlation time of 10^{-7} s was assumed for the protein and the complex, primarily to ensure efficient spin diffusion. The ligand correlation time was fixed at 0.2966 ns, corresponding to null NOE at 600 MHz. A leakage factor of $0.3\ s^{-1}$ was assumed for all the protons in their free and bound states. This term was added to all the diagonal elements of the R matrix.

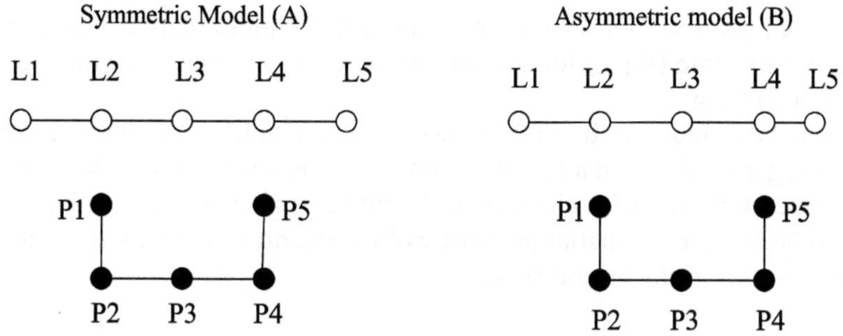

Fig. 2 **a** Symmetric model of ligand–protein complex. ***L1*** to ***L5*** represent the ligand, while ***P1*** to ***P5*** represent the protein. All the distances between the nearest neighbor protons in the complex are assumed to be 2.5 Å. **b** Asymmetric model, with the ***L4–L5*** distance set at 1.8 Å (to mimic the geminal protons)

4.1 Effect of Saturation Time on STD Intensities

Even if a given set of protein proton signals (***P2*** and ***P2′*** protons in Scheme 2) are instantaneously saturated, the saturation will take a finite time to spread to other protein protons (***P1*** and ***P1′*** sets of protons) and the bound ligand protons (***L′***) through dipolar networks, and through chemical exchange from the bound to the free ligand (***L***) protons. This is illustrated in Fig. 3, which

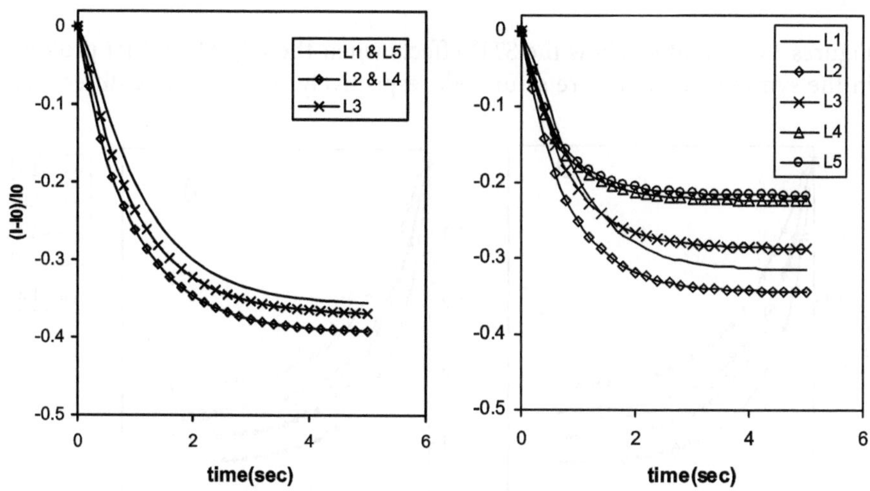

Fig. 3 **a** STD curves (*left panel*) for the symmetric ligand–protein complex in Fig. 2a. **b** STD curves (*right panel*) for the asymmetric ligand–protein complex in Fig. 2b. The ***P3′*** and ***P3*** protons are saturated. A spectrometer frequency of 600 MHz and a free ligand correlation time of 2.966×10^{-10} s corresponding to null NOE at 600 MHz were assumed. The protein correlation time was 10^{-7} s, Lt/Et = 10 : 1, leakage rate = $0.3\ s^{-1}$

shows the ligand STD-NMR fractional intensity changes plotted as a function of saturation time (Eq. 1) for the symmetric and asymmetric ligand–protein complexes in Fig. 2.

It may be noted that the STD intensities are changing even at 3 s of saturation, and generally reach a plateau in intensities at times greater than 4 s. The time-dependent intensity changes in the initial regions (< 0.1 s in Fig. 3) are more reflective of the spatial proximity of the ligand protons to the saturated protein proton in the bound state.

4.2 Effect of Ligand Relaxation Times and Conformation on STD Intensities

In Fig. 3b, ***L2*** and ***L4*** have substantially different STD values, with ***L4*** showing significantly smaller effect, *even though these two protons are equidistant* from the ***P3′*** proton. This is a simple consequence of the differences in the relaxation rates for these ***L2*** and ***L4*** protons due to differences in their "local environments" (e.g., in the asymmetric model, the ***L4–L5*** distance is shorter than the ***L3–L2*** distance, and thus ***L4–L5*** protons experience a faster longitudinal relaxation rate than the ***L4–L5*** protons). These observations suggest that caution is needed in qualitative attempts to relate the magnitudes of steady state STDs to spatial proximity of ligand protons to the protein protons.

4.3 Effect of the Relative Position of Saturated Proton(s) at the Active Site

Figures 3a, 4a and 4b show the STD effects when the ***P3***, ***P2***, and ***P4*** protons in the symmetric model are saturated, respectively. The dramatic difference

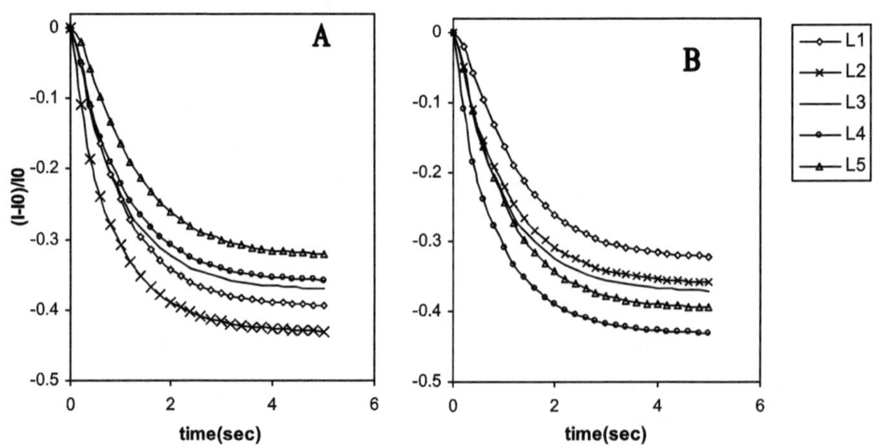

Fig. 4 Same as in Fig. 3a, but **a** with ***P2′*** and ***P2*** protons saturated and **b** with P4′ and ***P4*** protons saturated

in the magnitudes of the STD effects between the three figures is self-evident. This is a reflection of the fact that, even for long correlation times of 10^{-7} s with all the protein protons separated by 2.5 Å each (and a leakage of $0.3\ s^{-1}$), the saturation of protein protons due to spin diffusion is not 100% efficient in the complex as a result of *exchange-mediated leakage* [36, 37]. Thus, because of this exchange-mediated leakage phenomenon, saturation of protein protons adjacent to the bound ligand within the binding pocket generally produces larger STDs (due to direct effects, $P2' \rightarrow L'$) than saturation of protein protons remote from the binding pocket (due to the indirect effects, $P2' \rightarrow P1' \rightarrow L'$). Thus, some caution is needed in interpreting STD-NMR results based on the assumption of near 100% saturation of the entire protein resonance envelope due to irradiation at a single frequency.

4.4 Summary of the Properties of the STD-NMR Spectra

The simulations based on hypothetical models demonstrate the following factors:

1. The *apparent* epitope map based on STDs can exhibit rather significant dependence on which particular protein proton(s) is being saturated. For example, saturation of resolved methyl proton resonances from two separate residues may result in STD spectra with different relative and/or absolute intensities. Similarly, saturation of a specific tyrosine ring proton vs. a specific methyl group can result in different STD spectra.
2. Saturation of protein residues within the binding pocket is generally likely to produce larger STDs because of direct effects (i.e., $P2' \rightarrow L'$ in Scheme 2) than saturation of residues far away from the binding pocket, which transfer saturation through the indirect pathway ($P2' \rightarrow P1' \rightarrow L'$). This is a simple consequence of the exchange-mediated leakage [36, 37].
3. The STDs for small saturation times are independent of the free ligand conformation and its correlation time, and are more reflective of the bound-ligand proton distances with respect to saturated protein residue.
4. At longer saturation times the STDs exhibit a rather complex dependence on a number of factors including free ligand conformation and correlation time, and protein direct and indirect effects (Scheme 2). As a result, the relative intensities in the STD spectrum can sometimes vary with the saturation time.
5. The STDs and, hence, the *apparent* epitope map can exhibit significant dependence on intraligand relaxation rates, thus necessitating caution in simple qualitative interpretations. In particular, geminal protons on a ligand can show smaller STDs despite their proximity to the protein [36].

6. For very high ligand/protein ratios, the STD intensities are essentially independent of the K_d values in the range 10^{-3} M to 10^{-6} M [36].

5 CORCEMA Refinement of the Bound Ligand Conformation Within the Binding Pocket of the Target Protein Using STD-NMR Data: a Feasibility Study

Often, crystallographic structures and models generated by computational docking programs are used to generate bound-ligand structures within the binding pocket. However, the crystallographic structure may not be preserved in solution in its entirety (which affects the STDs significantly). Similarly, the docking programs may only generate approximate structures. Thus, it is essential to have a refinement tool that can take these structures and refine them quantitatively to generate models that are compatible with experimental STD-NMR data. We summarize below an approach by which such refinement can be obtained using STD-NMR intensities.

We developed an STD-NMR intensity-restrained CORCEMA optimization (SICO) procedure that utilizes a hybrid structure refinement protocol [49] involving a complete relaxation and conformational exchange matrix (CORCEMA) calculation [33, 35–37] and simulated annealing to refine the bound-conformation of a weakly binding ligand positioned within the binding pocket of a target protein. This method requires a knowledge of the PDB coordinates for the bound and free forms of the protein. This procedure is not designed to dock or orient a ligand within the binding pocket, or determine backbone conformation changes, if any, in the protein induced by the bound ligand; rather it is meant to find the global minimum conformation of the ligand giving the lowest NOE R-factor at a given docked position. In our approach, the global minimum for the bound-ligand conformation is obtained by a refinement of the torsion angles of an approximate (or starting) structure for the bound ligand using STD-NMR intensities as experimental constraints, and the NOE R-factor [45, 46] as the energy function to be minimized. The approximate structure for the complex may be obtained either from crystallographic data on the complex of interest, or generated from the crystallographic/NMR structure for a closely related complex, or from computer-docking of a ligand in the protein's binding pocket, or from NMR-based methods for docking [50, 51], or other approaches [52]. It also allows for an optimization of protein residue side-chains in the binding pocket. The flowchart for the CORCEMA-ST protocol with the implementation of simulated annealing (SA) [53–55] refinement for torsion angles and/or experimental parameters such as the bound ligand correlation time is shown in Fig. 1.

To unequivocally demonstrate the ability of our procedure in successfully identifying the global minimum conformation for the bound ligand, the most convincing way to demonstrate this method is using a STD data set where the global minimum (i.e., target) structure is known a priori so that one can objectively compare the results from CORCEMA-ST optimization of various starting structures with the target structure. The use of experimental STD data on a complex and its known crystallographic structure as the target structure will not meet this requirement since if there are differences in the CORCEMA-ST optimized structure and the crystallographic structure, it will be difficult to say if this is due to the crystallographic structures not being preserved in solution in their entirety, or if it is due to a shortcoming of our methodology. Thus, the best way to accomplish this objective is by utilizing a known crystallographic structure for the ligand–protein complex representing the target structure (global minimum), generate simulated experimental STD-NMR data for this target structure under typical experimental conditions (including noise in the 1D-NMR spectra represented by the elements of I_0), and show that the CORCEMA-ST procedure is able to *consistently and successfully* optimize any arbitrary starting conformation for the bound ligand to its known global minimum conformation.

For this purpose, we generated a hypothetical ligand–protein complex consisting of a trisaccharide (Gal(1,4)-[Fuc(1,3)]GlcNAc) using the PDB coordinates (ID# 2kmb) from the crystal structure of the complex sLe^x/MBP (sialyl Lewis^x tetrasaccharide/mannose binding protein) [56]. This complex is shown in Fig. 3 of [36]. We retained trisaccharide and eight protein residues (D184, E185, N187, H189, G190, N205, D206, and I207) in the MBP binding pocket for CORCEMA-ST calculations. Since STD-NMR measurements are usually performed in D_2O, we excluded all exchangeable hydrogens (OH and NH) in our calculations. For simulating the STD-NMR data, we assumed that the ring CH protons of His189 were instantaneously saturated, and the intensities were computed using Eq. 1 for a saturation time of 2 s, which is typical in many experiments. The ligand concentration was fixed at 0.5 mM, and calculations were performed at two separate ligand/protein ratios of 12.5 : 1 (protein ∼40 μM) and 300 : 1 (protein ∼1.67 μM). A uniform leakage relaxation of 0.2 s^{-1} was assumed for all the protons in their free and bound states, to mimic non-specific leakage relaxation with paramagnetic oxygen in the solution [33, 35]. The *simulated experimental* intensities ($S_{exp,k}$) were first calculated as percentage fractional intensity changes ($[(I_{0k} - I(t)_k) \times 100]/I_{0k}$), where k is a particular proton in the complex, and I_{0k} its thermal equilibrium value) from the intensity matrix $I(t)$ and using the target structure of the complex for the trisaccharide. Next, to demonstrate the ability of CORCEMA-ST program to successfully optimize the bound-ligand torsion angles to identify global minimum conformation (i.e., target conformation), we generated five different starting conformations for the bound ligand by *arbitrarily* changing the glycosidic torsion angles by significant amounts from those of the target structure.

For each starting structure and all subsequent intermediate structures at each cycle of SA refinement, the STD intensities of different protons in the ligand were calculated using Eq. 1 and compared with the *simulated experimental* STD intensities. This comparison utilized an NOE R-factor define as, [45, 46]:

$$\text{NOE R-factor} = \sqrt{\frac{\sum W_k (S_{exp,k} - S_{cal,k})^2}{\sum W_k (S_{exp,k})^2}} \,. \tag{6}$$

In these equations $S_{expt,k}$ and $S_{calc,k}$ refer to experimental and calculated STD values for proton k, and $W_k = 1/S_{exp,k}$. The use of a weighting (W_k) proportional to $1/S_{exp,k}$ for each individual STD intensity has the effect of making the R-factor sensitive to significant deviations in small STD values as well as to deviations in the large STD values. The NOE R-factor (Eq. 6) is taken as the pseudo-energy function to be minimized in a combined CORCEMA/SA refinement. Since occasionally there could be steric clashes during optimization, the CORCEMA-ST program has a provision to add, if necessary, an empirical van der Waals repulsion term [57] to Eq. 6 to minimize the conflicts between atoms during the refinement [49]. This term, E_{repel}, is defined by [49, 57]:

$$\begin{aligned} E_{repel} &= 0 && \text{if} \quad r \geq s r_{min} \\ & k_{vdw}(s^2 r_{min}^2 - r^2)^2 && \text{if} \quad r < s r_{min} \,. \end{aligned} \tag{7}$$

If necessary, the optimized structures can be subjected to additional energy minimization separately (outside CORCEMA-ST) using standard force field parameters in programs such as AMBER or CHARMM.

In the ideal case of a successful global energy minimization using STDs without any experimental errors, the NOE R-factor is zero. In practice however, the STD intensities can have significant random errors depending upon the S/N of the 1D-NMR spectra with (I) and without (I_0) rf irradiation, from which the difference NMR spectrum ($I - I_0$) is obtained. Thus, in practice the NOE R-factor will always be greater than zero even when the global minimum is achieved during optimization. To test the ability of the CORCEMA-ST program to optimize structures using more realistic STD data, we generated simulated data sets including random noise. For this, we first generated **I** and $\mathbf{I_0}$ matrices for the crystallographic structure of the target structure, and then added 1%, 2.5%, or 5% of the total ligand concentration (0.5 mM) multiplied by a random number between −1 and +1 to the elements representing ligand proton intensities in the column matrices **I** and $\mathbf{I_0}$, separately. Using this procedure, random noise levels of 1%, 2.5%, and 5% in the 1D-NMR spectrum correspond respectively to S/N ratios of 125 : 1, 50 : 1, and 25 : 1 in the 1D-NMR spectra (i.e., I_0 elements representing the ligand).

For the results shown here, the free ligand conformation for the carbohydrate is maintained to be that in the crystal structure. We also assumed that the protein conformation is known (e.g., from crystallography or NMR) and is fixed to be identical in both the free and bound states (i.e., there are no ligand-induced backbone conformational changes). Note that the side chain orientations of the protein residues within the binding pocket could also be optimized using our procedure. In particular, this may be necessary if the optimization of ligand structure alone doesn't yield acceptable R-factors in spite of using high-quality STD data sets. The PDB coordinates of the free ligand, free protein, and protein–ligand complex are read-in parameters. The remaining parameters such as correlation times and binding constants were also fixed, as described below.

In order to obtain meaningful optimization, it is preferable for the number of parameters to be optimized to be less than the number of experimental STDs. In the current set of calculations, in addition to the four ligand torsion angles, four additional parameters are needed: the dissociation constant (K_d), the correlation times for the free ligand (τ_L) and the protein (τ_P), and the order parameter S^2 for methyl group-external proton interactions [36, 41, 42]. Since the STDs are independent of K_d in the range 10^{-4} to 10^{-7} M for high ligand/protein ratios [36], we chose arbitrarily a typical value of 10^{-6} M for the weak complex to compute the exchange matrices using a diffusion limited to a rate of 10^8 s^{-1} M^{-1}. The NOE R-factor is not very sensitive to variations in S^2 in the range 0.6–0.9 (changing only by about 0.05 in magnitude, results not shown); thus a reasonable value of 0.85 was assumed for S^2. The τ_L can always be experimentally determined; here we chose a typical value of 0.5 ns for the trisaccharide. Thus, the only remaining parameter needed is τ_P. This correlation time can be determined experimentally from independent methods (e.g., from fluorescence decay). Alternatively, this could be determined by simultaneous optimization together with the torsion angles. For the current calculation τ_p was fixed at 40 ns. A spectrometer frequency of 600 MHz was assumed. For the carbohydrate ligand, the sampling range for the interglycosidic torsion angles ϕ: H1–C1–O–Cn (–60° to + 120°) and for ψ: C1–O–Cn–Hn (–60° to +60°) where Cn and Hn refer to the aglycon carbon and its attached proton (Fig. 5), were restricted to the allowed regions as per the Ramachandran- type population contour map of the glycosidic linkages [58] in sLe^x. Each ϕ (across C1–O) and ψ (across O–Cn) optimization was done in both directions of the pertinent bonds as the intermolecular dipolar interactions, and hence the associated STDs, are different for each choice. For peptide ligands, the sampling range for the torsion angles is from –180° to +180° [49].

Figure 6a shows a comparison of the five starting structures for the trisaccharide and the target structure. Figure 6b shows the results of optimization of the five starting structures using simulated STD data sets (with 23 intensities) without any noise. They are shown within the protein binding pocket, which is included in the calculation. The torsion angles and the NOE R-factors for this case before and after CORCEMA-ST optimization are summarized

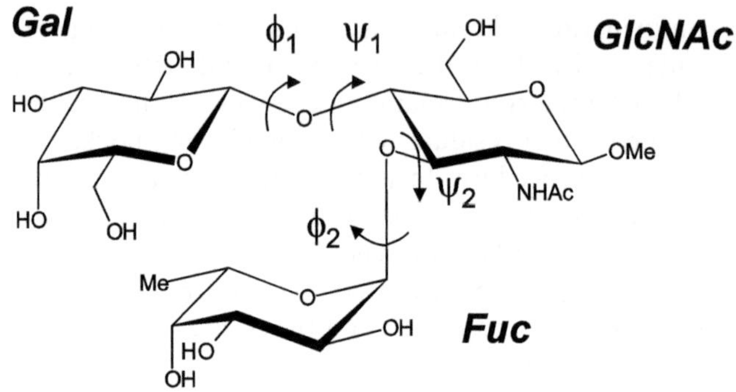

Fig. 5 Structure of the trisaccharide (Gal(1,4)-[Fuc(1,3)]GlcNAc)

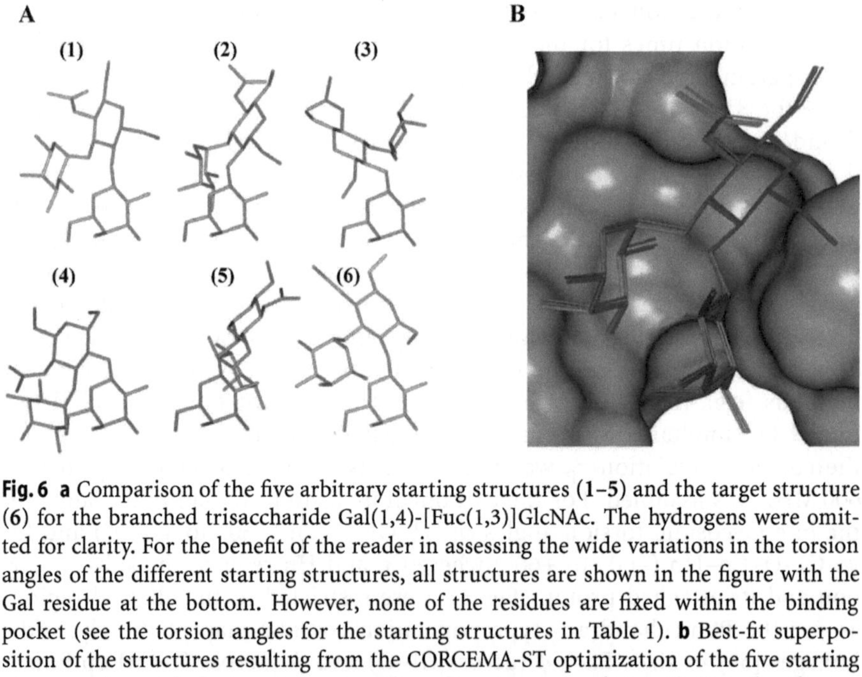

Fig. 6 **a** Comparison of the five arbitrary starting structures (**1–5**) and the target structure (6) for the branched trisaccharide Gal(1,4)-[Fuc(1,3)]GlcNAc. The hydrogens were omitted for clarity. For the benefit of the reader in assessing the wide variations in the torsion angles of the different starting structures, all structures are shown in the figure with the Gal residue at the bottom. However, none of the residues are fixed within the binding pocket (see the torsion angles for the starting structures in Table 1). **b** Best-fit superposition of the structures resulting from the CORCEMA-ST optimization of the five starting structures in **a** with the target structure (ligand/protein ratio of 12.5 : 1). Note that the target structure is shown at a slightly different orientation than the one in **a**. The noise level was assumed to be zero. The smooth surface represents the protein residues within the binding pocket. Table 2 shows the results of successful optimization when the 1D-NMR spectra (i.e., I_0) are recorded with a finite S/N value prior to calculating STDs. Reprinted from [49]; © 2004, with permission from Elsevier

in Table 1. It is seen that CORCEMA-ST is successfully able to optimize all the five different structures to the same global minimum structure (the target structure). We obtained equally good convergence when we reduced the

Table 1 Results of optimization without noise of different starting structures for the trisaccharide (ligand/protein ratio = 12.5 : 1)

Ligand conformation	$\phi 1$	$\psi 1$	$\phi 2$	$\psi 2$	R-factor
Target (crystal) structure	*44.32*	*14.88*	*49.77*	*20.69*	*0.00*
Starting structure 1	14.28	- 35.11	99.74	50.70	2.10
Optimized structure 1	45.13	14.91	48.57	19.77	0.02
Starting structure 2	- 25.67	- 20.20	109.81	- 14.45	2.19
Optimized structure 2	43.22	14.57	47.39	22.74	0.02
Starting structure 3	107.11	58.23	- 54.32	- 46.09	0.82
Optimized structure 3	43.40	14.68	49.27	20.72	0.02
Starting structure 4	114.23	- 55.16	- 49.82	55.73	0.70
Optimized structure 4	44.09	14.55	49.79	20.18	0.02
Starting structure 5	- 45.72	- 55.04	- 45.26	- 49.16	6.62
Optimized structure 5	40.95	17.39	48.49	20.25	0.02

Table 2 Results for the optimization of the same starting structure (#2 in Table 1) for the trisaccharide using simulated STD-NMR data obtained from 1D-NMR spectra with different S/N values (ligand/protein ratio = 12.5 : 1)

Ligand conformation	S/N[a] of 1D-NMR spectra	$\phi 1$	$\psi 1$	$\phi 2$	$\psi 2$	R-factor
Target structure	–	*44.32*	*14.88*	*49.77*	*20.69*	*0.00*
Initial structure 2	–	- 25.67	- 20.20	109.81	- 14.45	2.19
Optimized structure 2	High	43.22	14.57	47.39	22.74	0.02
Optimized structure 2A	125 : 1	45.85	14.37	54.77	18.98	0.05
Optimized structure 2B	50 : 1	40.34	17.43	49.08	21.72	0.07
Optimized structure 2C	25 : 1	58.65	6.99	29.98	31.02	0.18

[a] S/N here refers to the normal 1D-NMR spectrum of the ligand, i.e., the elements of I_0, and *not* the STD-spectrum itself, which is the difference $I–I_0$. The corresponding STD spectra ($I–I_0$) will have substantially lower S/N. See the text for details on adding noise to elements of I and I_0

number of STDs included in the calculation from 23 to 11. In Table 2, we have summarized the results of systematically varying the S/N ratio in the 1D-NMR spectra (Note: S/N here refers to the normal 1D-NMR spectrum of the ligand, i.e., the elements of I_0 and *not* the STD-spectrum which is the difference $I–I_0$) on the ability of the CORCEMA-ST program to locate the global minimum at a ligand/protein ratio of 12.5 : 1. It is seen that at 12.5 : 1 ratio, good convergence is obtained even when the S/N ratio is only 50 : 1 for the 1D-NMR spectrum.

Similar calculations at a ligand/protein ratio of 300 : 1 were also performed. But at 300 : 1 ratio a reasonable convergence is obtained if the 1D-NMR spectra (I_0) are recorded at an S/N ratio of 1250 : 1 or better [49], which is very routinely accomplished these days with the use of perdeuterated solvents, extensive signal averaging, and high magnetic fields. As the number of optimizable parameters increases (e.g., the number of torsion angles), recording 1D-NMR spectra with better S/N ratios is necessary to avoid getting trapped in local minima. The SICO protocol that we developed belongs to the general class of intensity-restrained structure optimization procedures described previously in the literature [45, 46, 59–62], but incorporates the reversible exchange of the interacting molecules between their free and complexed states in an explicit manner.

5.1 Calculations Involving Proteins Without Assignments

To undertake CORCEMA calculations, it is necessary to set up the **Q**-matrix, which requires a knowledge of the assignment of the protein resonance(s) being saturated. Thus, in those instances where the proteins have already been extensively characterized and the NMR assignments completed, this method readily enables one to refine the bound structures of weakly binding ligands using STD data. In practice, however, many STD-NMR measurements involve novel target proteins for which the assignments are not available. Such unassigned proteins may potentially pose a challenge for undertaking these calculations. For such proteins, we suggest the following practical approaches in generating usable STD data for undertaking CORCEMA-ST calculations:

1. Selective saturation (using shaped pulses) of all the aromatic ring CH resonances together since they show up generally in a short window of 6.5–8.5 ppm. The **Q**-matrix is easily set up for this case.
2. Selective saturation of well-resolved ring CH protons of an easily identifiable residue such as Tyr or His. If this residue happens to be unique (i.e., appearing only once in the sequence), its assignment is known. If there is more than one Tyr or one His in the sequence, it may be still possible to infer its assignment because if these residues are part of the protein binding pocket they tend to produce larger STDs (due to direct effects, $E2' \rightarrow L'$) at low to moderate ligand/protein ratios (15 : 1 to 50 : 1). If they are far away (more than 7 Å) from the ligand, the STDs are much smaller [36].
3. Similarly, the methyl resonances which appear as a group (generally between −1 ppm and 1.5 ppm) can be saturated together and incorporated in

the **Q**-matrix calculation. If some non-methyl proton resonances (e.g., Lys γCH_2) happen to overlap with this methyl group, errors introduced by their saturation are small provided these protons are well outside a 7 Å distance from the ligand in the binding pocket. If these non-methyl resonances arise from residues within the pocket or in its vicinity, there will be errors.

4. On the other hand, if the saturation of a well-resolved methyl resonance produces large STDs at moderate ligand/protein ratios, it may be possible to infer its assignment by looking for methyl-group containing residues in the binding pocket. If there are two or more such methyl-group residues either within the protein binding pocket or within a 7 Å cut-off from the ligand, it may be possible to infer the assignments indirectly from the magnitudes of optimized NOE R-factors.
5. Finally, a saturation-transfer screening (using short saturation times) of the protein resonance envelope to identify irradiation frequencies that yield large STDs may be of some aid in the identification of residues within the binding pocket, and in assignments for the calculations.

6 Quantitative Analysis of Experimental STD-NMR Data Using CORCEMA-ST Calculations

6.1 Sialoadhesin–Sialyl Lactose Complex

The siglecs are a family of I-type lectins binding to sialic acids on the cell surface. Sialoadhesin (siglec-1) is expressed at much higher levels in inflammatory macrophages and specifically binds to α-2,3-sialylated *N*-acetyl lactosamine residues of glycan chains. The terminal disaccharide α-D-Neu5Ac-(2 → 3)-β-D-Gal is thought to be the main epitope recognized by sialoadhesin. To understand the basis of this biological recognition reaction, STD-NMR experiments were employed to characterize the binding epitope of α-2,3-sialylated lactose, α-D-Neu5Ac-(2 → 3)-β-D-Gal-(1 → 4)-D-Glc to sialoadhesin at atomic resolution using quantitative analysis of STD-NMR data using the CORCEMA-ST method [63]. The experimental STD intensities were compared to the theoretical STD values predicted for the crystal structure of sialyl lactose and sialoadhesin complex [64] using CORCEMA-ST procedure.

The STD-NMR experiments were performed on a Bruker Avance DRX 500 spectrometer at 303 K. The NMR samples were prepared in D_2O with the ligand/protein ratios of 1000 : 1, 500 : 1, 300 : 1, and 100 : 1 ([E] =5 μM). Other experimental details are contained in [63].

Calculations were performed using the crystallographic structure of the sialyl lactose–sialoadhesin complex, PDB crystal structure entry 1QFO [64].

In their uncomplexed states, for simplicity in calculations, the protein and the ligand were assumed to retain the same conformation as in the complex. In the calculations, three sugars (Sia201 (Neu5Ac), Gal202, Glc203) and the twenty amino acid residues within the binding pocket (Trp2, Thr37, Ala38, Ile39, Tyr41, Tyr44, Ser45, Arg48, Arg97, Phe98, Glu99, Ile100, Ser101, Ser103, Asn104, Arg105, Trp106, Leu107, Asp108 and Val109) were included.

Since the protein signals at 0 ppm were irradiated for the STD experiment, we made the reasonable assumption that the Leu, Val and Ile methyl protons were instantaneously saturated. From the intensity matrix $\mathbf{I}(t)$, the fractional intensity changes for different ligand protons k were calculated, and compared to the experimental STD values using an NOE R-factor Eq. 6. The value of W_k was assumed to be 1.

The multilevel coordinate search (MCS) method was used to optimize parameters to get best fit between the experimental and predicted intensities. The NOE R-factor was used as the energy function to be minimized. A version of the MCS method was written based on the version presented by Huyer and Neumaier [65]. The algorithm performs the minimization by a standard coordinate search method. The method was carefully tested by the use of different starting points for the coordinate search and using simulated data sets. This alleviates local minima trapping by MCS, and identifies the global minimum within the parameter ranges used in the optimization.

The dissociation constant (K_d) of the sialyl lactose–sialoadhesin complex was reported as 0.8 mM from the changes in chemical shift of sugar resonances and 1.4 ± 0.4 mM from the changes in chemical shift of protein resonances [66]. Thus the calculations were performed by restricting the K_d values to the range 0.8–1.4 mM. The methyl group internal correlation times (τ_m) was set at a reasonable value of 5 ps [67]. The k_{on} was set to $10^8\ s^{-1}\ M^{-1}$ and the leakage factor was set at $0.1\ s^{-1}$. For optimization of the parameters the following ranges were employed initially on the 300 : 1 ratio data set: $\tau_{r\text{-protein}}$ (10^{-8}–10^{-7} s), $\tau_{r\text{-ligand}}$ (0.1–2 ns), external methyl S^2 (0.6–0.85), and K_d (0.8–1.4 mM) at 1 : 300 ratio of protein (5 μM) to ligand. The calculations were repeated with the narrow ranges $\tau_{r\text{-protein}}$ (2×10^{-8} s to 6×10^{-8} s), $\tau_{r\text{-ligand}}$ (1–2 ns), external methyl S^2 (0.7–0.85), and K_d (0.8–1.4 mM) for the same ratio of protein to ligand and found that it converges to the same global minimum.

To speed up the computational time, we restricted further calculations within the above narrow ranges of parameters for the remaining protein/ligand ratios. In all the calculations the protein concentration was kept fixed at 5 μM. The predicted STDs in Fig. 7 were calculated using the following parameters corresponding to global minimum: $\tau_{r\text{-ligand}} = 2$ ns; $\tau_{r\text{-protein}} = 3 \times 10^{-8}$ s; $k_{on} = 10^8\ M^{-1}\ s^{-1}$; uniform leakage = $0.1\ s^{-1}$; $S1^2 = 0.85$; and $K_d = 1.4$ mM. The B-chain of the crystal structure was used in these calculations.

From the CORCEMA-STD predictions based on the crystal structure, all protons in *N*-acetyl neuraminic acid and galactose show reasonable STD

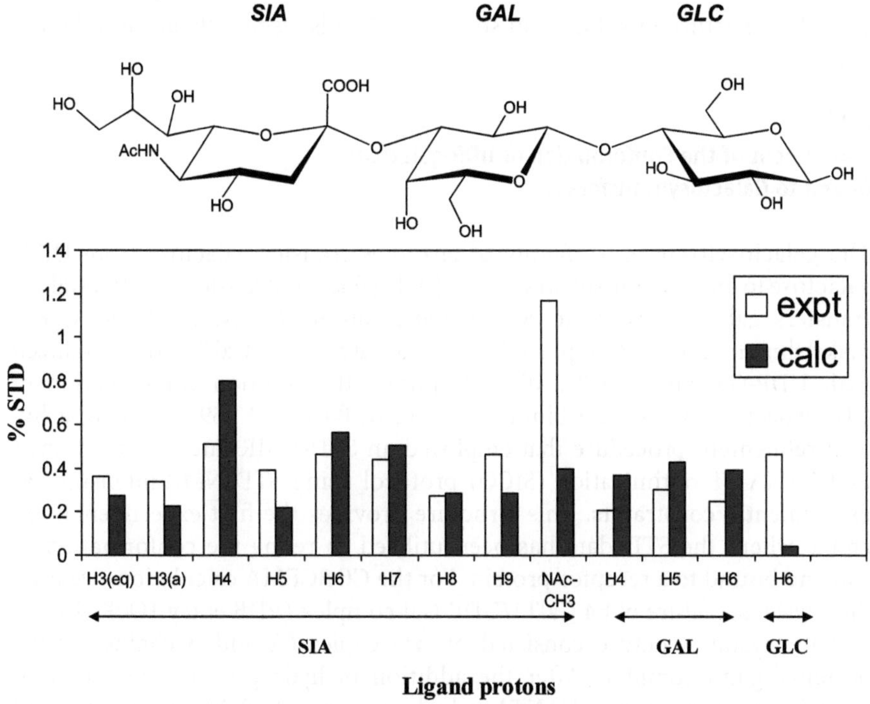

Fig. 7 Comparison of experimental and predicted STD values from CORCEMA-ST protocol for the crystal structure of sialoadhesin–sialyl lactose complex at protein/ligand ratio of 1 : 300. Reprinted from [63]; © 2004, with permission from Elsevier

magnitudes, indicating that these two residues are in intimate contact with the protein. In contrast, the protons of Glc show negligibly small STDs suggesting that this residue is somewhat away from the protein. If we compare the calculated and experimental STD values of individual protons, the calculated values are significantly smaller than the experimental values for the *N*-acetyl methyl group of *N*-acetyl neuraminic acid and for H6 of glucose. This is not surprising since in the crystal structure the *N*-acetyl methyl group is too far from the protein methyl groups to experience a "direct" saturation transfer. The R-factors (0.57–0.67) calculated at different protein/ligand ratios are in general indicative of a moderate overall fit between experimental and calculated values. Exclusion of STD values for the *N*-acetyl methyl group and H6 of glucose resulted in R-factors ranging from 0.4 (at 1 : 300) to 0.49 (at 1 : 500), leading to a slightly improved fit.

Considering an estimated experimental error of 10–20% for the STD data as well as some systematic errors in the estimates from resonance overlap, the analysis suggests a generally reasonable correspondence between the crystal structure and the solution state. At the same time, our data suggest that the crystallographic structure of the complex may not be entirely compatible with the structure in solution in some fine details, especially around the Glc residue.

6.2 Refinement of the Conformation of UDP-galactose Bound to Galactosyltransferase

The galactosyltransferase family of enzymes transfer galactose from UDP-galactose to an acceptor substrate, e.g., β1,4-galactosyltransferase-I (β4 GalT1) transfers galactose (Gal) to *N*-acetylglucosamine (GlcNAc). Recently, Ramakrishnan et al. [68] reported the structure of β4 GalT1 co-crystallized with UDP-Gal and $MnCl_2$. We determined the solution conformation of UDP-galactose within the binding pocket of β4 GalT1 [69, 70] using a hybrid refinement procedure that employed an STD-NMR intensity-restrained CORCEMA-ST optimization (SICO) protocol using STD NMR-intensities as experimental constraints. This structure provides the first experimental example where the STD data has been utilized to refine the conformation of a ligand bound to a receptor protein. For the CORCEMA calculations we used the crystal structure of β4 GalT1/UDP-Gal complex (PDB entry 1O0R) [68].

The crystal structure consisted of two copies (A and B chains) of the protein–ligand complex. After the addition of hydrogens to the crystallographic structure using QUANTA, a bad contact (~1.39 Å) between H3 and H5 protons of the ribose residue was observed in both A and B chains. Further, the O5–C5–C4 bond angle in the ribose (139.5°) showed a significant deviation from the normal tetrahedral angle [71]. To fix this bad contact and the abnormal bond angle observed in the crystal structure, we performed an energy minimization of the complex (using Insight-II) by only including the residues within the binding pocket. In the energy-minimized structure the O5–C5–C4 bond angle in ribose is 112.6 and the bond distance between H3 and H5 protons of ribose is 2.59 Å. Thus, we have used this energy-minimized crystal structure as the starting structure for the complex in CORCEMA optimization. In the uncomplexed state, the protein structure was defined by the coordinates for the free enzyme [72]. The free ligand was assumed to have the same torsion angles as in free UDP-glucose, which is very similar [73]. The UDP-galactose ligand and 27 protein residues within the binding pocket (I186, P187, F188, R189, N190, R191, Q192, H194, F226, R228, S251, D252, V253, D254, L255, K279, Y289, G292, G313, W314, G315, E317, D318, M344, H347, D350 and N353) were included in the CORCEMA calculations. The experimental STD-NMR data recorded on UDP-Gal at 500 MHz at a ligand/protein ratio of 55 : 1, 293 K and a saturation time of 2 s using the procedures described before [74] were utilized in these calculations.

Since the protein signals at zero ppm were saturated in the STD experiment, for the computation of the Q-matrix in Eq. 1 we made the assumption [70] that the I186, V253, and L255 methyl protons are instantaneously saturated. This results in a very high R-factor (1.85 for the crystal structure and 1.24 for the energy-minimized crystal structure). This is not surprising since in the crystal structure the methyl group (pro-R) of V253 is very close (~2 Å) to the H2 proton of the ribose residue, thus resulting in a significantly larger STD value for the ribose H2/H3 protons compared to the experiment. After careful examination of the crystal structure, we made the reasonable assumption that the V253 methyls escape saturation (due to low-field shifts induced by proximal aromatic rings). This results in a considerable improvement in the R-factor (0.587 for the crystal structure and 0.428 for the energy-minimized crystal structure), but still with significant deviations for H2/H3R and H5U. However, optimization of the energy-minimized crystal structure using STD-NMR intensity-restrained CORCEMA optimization (SICO) resulted in a substantial improvement in the NOE R-factor (0.306) and in the overall fit between calculated and experimental data.

In our calculations, in addition to the torsion angles, four additional parameters are needed: the dissociation constant (K_d), the correlation times for the free ligand (τ_L) and the protein (τ_P), and the order parameter S^2 for methyl group-external proton interactions. The dissociation constant value of UDP-Gal in complex with β4 GalT1 was measured to be 1.6 μM at 298 K. We have assumed the diffusion limited on rate as 10^8 s^{-1} M^{-1}. The NOE R-factor is not very sensitive to variations in S^2 in the range 0.6–0.9; thus a reasonable value of 0.8 was assumed for S^2. For τ_L here we chose a typical value of 0.5 ns for the ligand, based on its low molecular weight. A uniform leakage relaxation of 0.2 s^{-1} was assumed for all the protons in their free and bound states. The methyl group internal correlation time (τ_m) was set at a reasonable value of 10 ps. For the free ligand, the torsion angles were taken from [73] ($\alpha = 60°$, $\beta = -180°$, $\gamma = 59°$, $\phi = 74°$, $\psi = -90°$, $\chi = -178°$, $\mu = -179°$, $\nu = 90°$). Thus, the only remaining parameter needed is τ_P, this was determined by simultaneous optimization together with the torsion angles. During the optimization we fixed the torsion angles across bonds linking the phosphate groups since they are known to be coordinated with the metal ion in the crystal structure [71]. Thus the SICO procedure included the "simultaneous" refinement of six torsion angles (galactose (ϕ, ψ, and ω) and ribose (α, β and γ)) in the ligand (Fig. 8), and the protein correlation time. The galactose orientation is defined by the torsion angles ϕ (P2–O1″-C1″-O5″), ψ (OP4–P2–O1″–C1″) and ω (O5″–C5″–C6″–O6″), and the ribose orientation is defined by α (Op4–P1–OP1–C5′), β (P1–OP1–C5′–C4′) and γ (OP1–C5′–C4′–C3′). The numbering of the atoms is same as Scheme 1 in [73]. The sampling range chosen for the interglycosidic angles was from −180° to +180° and for the protein correlation time 0.1–100 ns. The target function optimized was a sum of the NOE R-factor and the van der Waals repulsion term. Table 3 shows a comparison of the experimental and calculated STDs for the various models. The NOE R-factors for different structures are shown at the bottom.

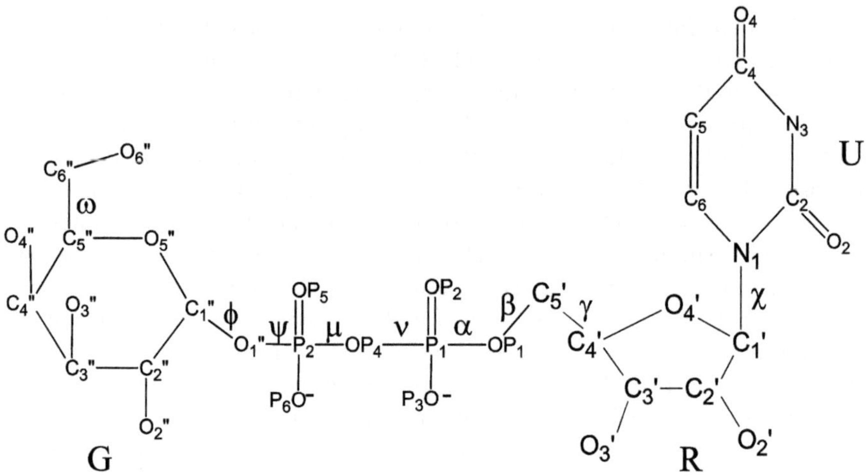

Fig. 8 Structure of UDP-Gal with labeling of the residues (*G* galactose, *R* ribose, *U* uracil) and torsion angles. Reprinted with permission from [69] (supporting information); © 2004, American Chemical Society

Table 3 Comparison of experimental and calculated percentage fractional STD intensities for UDP-Gal/GalT1 complex

Proton	Experimental STDs	STDs from CORCEMA-ST method Crystal structure	Energy minimized crystal structure	SICO structure
H5 U	3.1[a]	0.7095	0.9049	1.2653
H6 U	0.96	1.0961	1.1262	1.3414
H1 R	6.4	8.2537	5.5857	6.5582
H2/H3 R	2.6	5.1683	4.6646	2.9718
H4 R	2.3	2.8104	1.6600	1.4764
H5/H5′ R	1.2	2.3522	0.9158	0.8532
H1 G	0.73	1.0030	0.8599	0.7957
H2 G	1.1	1.6217	1.1691	1.2470
H3 G	0.99	1.4968	1.0964	1.2203
H4 G	1.1	1.8003	1.4872	1.4592
H5 G	0.76	0.4757	0.4276	0.4822
R-factor		**0.587**	**0.428**	**0.306**

Protein correlation time = 23.5 ns; $K_d = 1.6 \times 10^{-6}$ M; time delay = 2.3 s; $t = 2$ s; Order parameter = 0.8; ligand correlation time = 0.5 ns; Rho leak = 0.2 s^{-1}; $\tau_m = 10$ ps; [L] = 0.8 mM; [E] = 14.5×10^{-6} M

[a] H5U has a systematic error due to resonance overlap

Figure 9 shows a comparison of the energy-minimized structure and the CORCEMSA-ST structure for the ligand, along with the protein residues in the binding pocket. The discrepancy between experimental and SICO-predicted STDs for the H5U may be the result of systematic error from the overlap of H5U and H1R resonances [74]. The UDP-Gal torsion angles are shown in Table 4 for the crystal structure, the energy-minimized crystal structure, and the CORCEMA-ST refined structure.

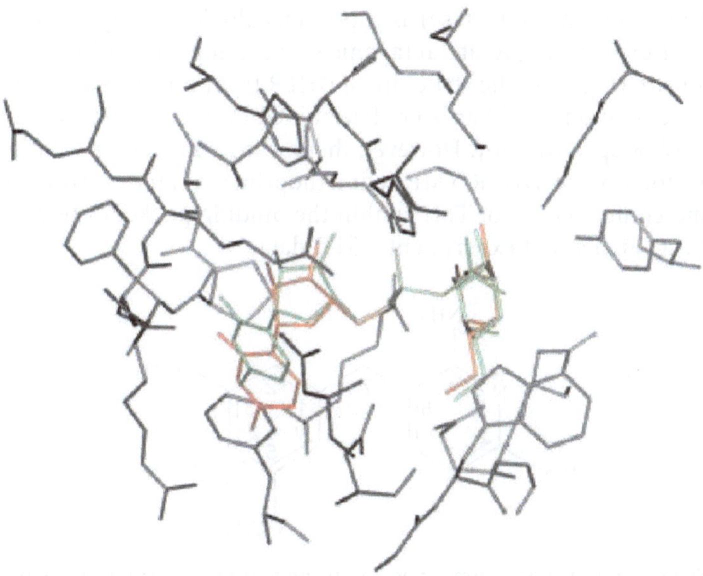

Fig. 9 Best-fit superposition of CORCEMA-ST optimized structure (*red*) with the energy-minimized crystal structure (protein residues within the binding pocket are *black* and ligand residues are *green*). The hydrogens were omitted for clarity. Reprinted with permission from [70]; © 2005, American Chemical Society

Table 4 Comparison of torsion angles for different structures of UDP-Gal/β4 GalT1 complex

Torsion angle	Crystal structure	Energy-minimized crystal structure	SICO structure
Ribose (α)	- 146.22	149.97	145.51
Ribose (β)	- 177.11	106.06	103.73
Ribose (γ)	164.22	- 77.94	- 91.60
Galactose (ϕ)	73.55	75.84	83.60
Galactose (ψ)	- 133.04	- 132.85	- 131.88
Galactose (side-chain)	172.39	161.1	130.85

In conclusion, using high-quality STD-NMR data on UDP-Gal weakly binding to β4 GalT1, we demonstrated that it is possible to refine the crystal structure (or any computer-docked structure in the proper orientation that serves as the starting structure) to obtain a global-minimum conformation for the bound ligand in solution.

6.3 Dihydrofolate Reductase (DHFR)/Trimethoprim (TMP) Complex

DHFR (dihydrofolate reductase) is a pharmacologically important intracellular target enzyme for folate antagonists, including the antibacterial agent trimethoprim (Fig. 10). The structure of DHFR from various sources with and without the bound ligand has been determined by X-ray crystallography and solution NMR spectroscopy. However, there was no crystal or solution NMR structure for bovine DHFR/TMP (trimethoprim) complex. We determined the bound conformation of TMP within the binding pocket of bovine DHFR from a SICO analysis of experimental STD data.

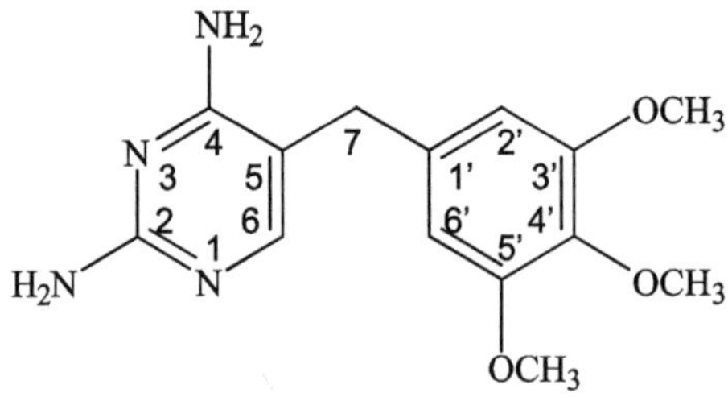

Fig. 10 Structure and atom numbering for trimethoprim. Reprinted with permission from [75]; © 2005, American Chemical Society

The starting structure for the bovine DHFR/TMP complex was modeled from the crystal structure [76] of chicken liver DHFR-TMP-NADPH ternary complex (coordinates received from Dr. D. A. Matthews). The Biopolymer module in Insight-II (Accelrys) was used to make the necessary amino acid changes in the chicken liver DHFR (Y31F and K32Q) in order to generate the bovine DHFR complex. We performed an energy minimization of the substituted residue in the bovine DHFR complex by using the Discover module in Insight-II using the steepest descent algorithm and the force field CVFF. This energy-minimized bovine DHFR/TMP complex was used for STD predictions. In the current calculations 35 amino acid residues within a 7 Å distance from the ligand were included (I7, V8, A9, V10, I16, L22, W24, L27, E30, F31, Q32, Y33, F34, Q35, T38, V50, I51, M52, T56, W57, S59, I60, P61, N64, L67, R70, I114, V115, G116, Y121, F134, V135, T136, I138, F179). The experimental STD data were directly taken from Fig. 2 of [77]. We selected STD data sets at nine different saturation times (0.9, 1, 1.3, 1.6, 2, 2.6, 3, 3.5 and

4 s). The STD data in Fig. 2 of [77] were obtained on a 600 MHz spectrometer with a ligand/protein ratio of 35 : 1 ([E] =120 μM). The on-resonance saturation frequency was 0.9 ppm whereas the off-resonance irradiation frequency for the control spectrum was at −9.5 ppm. The total recovery delay was set to 4 s. Since STD-NMR measurements were performed in D_2O, we have excluded all exchangeable hydrogens (OH and NH) in our calculations.

In our calculations, the parameters needed for STD predictions are the dissociation constant (K_d), the correlation times for the free ligand (τ_L) and the protein (τ_P), the methyl group internal correlation time (τ_m), the leakage relaxation for all the protons in their free and bound states (to mimic non-specific leakage relaxation with paramagnetic oxygen in the solution), and the order parameter S^2 for methyl group–external proton interactions [41, 42]. The association constant value of TMP in complex with DHFR is set at 2×10^7 as determined previously [78]. We have assumed the diffusion limited on rate as $10^8\ s^{-1}\ M^{-1}$. A reasonable value of 0.85 was assumed for S^2. Thus, the remaining parameters needed are τ_P, τ_L, τ_m, and the leakage relaxation. These parameters were determined by a SICO protocol [49] using the experimental STD-NMR intensities as the constraints. In parameter optimization, the NOE R-factor was used as the target function. In torsion angle refinement, the target function that was optimized was a sum of the NOE R-factor and a van der Waals repulsion term [69, 70], which avoids the conflicts between atoms during the torsion angle refinement. The simulated annealing optimizations were performed by fitting the experimental STD growth curves composed of nine different saturation times for each of the five different resonances in the TMP ligand. This resulted in 45 STD intensities as experimental constraints and all these STD values were used in the calculations of NOE R-factors.

The free ligand conformation was assumed to be same as the bound ligand conformation in the chicken liver DHFR complex, as suggested by others previously. Since the protein signals at 0.9 ppm were saturated in the STD experiment, for the computation of the Q-matrix in Eq. 1 we made the reasonable assumption that all protein protons with chemical shift values within 0.4 ppm of the saturation frequency (0.9 ppm) were instantaneously saturated. These included the methyls of V8, A9, V10, L22, L27(D2), L67, I114, V115, V135, T136 as well as the β hydrogens of I7, L22(HB2), I60. For the chemical shifts of DHFR we utilized the published assignments of homologous human DHFR/methotrexate binary complex [79]. TMP molecule (Fig. 10) has three methoxy groups, two of which (3′ and 5′ methoxy groups) are symmetrically located. Hence, we assumed two different internal methyl correlation times for the methyl groups in the free ligand (τ_{m1} for 4′Ome and τ_{m2} for 3′/5′Ome). An independent calculation showed that the bound internal methyl correlation time has a negligible effect on the STD values. Thus, for simplicity we assumed only one internal methyl correlation time (τ_{m3}) for the bound ligand and enzyme methyl groups. This makes six parameters to be optimized.

The sampling range for the τ_L is 0.1–1 ns, for the τ_P is 0.1–100 ns, for the τ_m is 0.5–10 ps and for the leakage relaxation is 0.05–0.3 s^{-1}. The optimized NOE R-factor for the initial model of bovine DHFR/TMP complex was 0.076 and the optimized parameters were: $\tau_L = 0.101$ ns; $\tau_P = 20.43$ ns; $\tau_{m1} = 0.81$ ps; $\tau_{m2} = 3.04$ ps; $\tau_{m3} = 3.26$ ps; and the leakage factor = 0.065 s^{-1}. The resulting STD buildup curves observed for the six parameter optimization condition are shown in Fig. 11. The low R-factor and the reasonable values of all the optimized parameters from the wide sampling range reveal the quality of the optimization. The utilization of a large data set (containing 45 STD values) including STDs during the growth portion imposed considerable demands, and it is heartening that the SICO optimization was very successful.

This very low NOE R-factor suggests that the predicted STD values for the initial model of the bovine DHFR/TMP complex generated using the crystal structure of chicken liver DHFR/TMP as a template are in excellent agreement with the experimental STD intensities measured for the bovine DHFR/TMP complex in solution. However, a careful examination of the STD values measured for various protons in TMP suggests that the 7-CH_2 geminal protons and 4′OMe methyl signal shows some systematic deviations from the experimental STD values. For the 7-CH_2 geminal protons the predicted STD values are uniformly higher than the experimental values, whereas for the 4′OMe methyl STDs increasingly larger deviations between experimental and predicted values are observed at longer saturation times. When we carefully examined the crystal structure we found that 7-CH_2 geminal proton experiences direct saturation from its nearest neighbor (< 3 Å) L22 methyl groups. The chemical shift value of the L22 methyl groups are 0.51 and 0.8 ppm. In the STD experiment protein is irradiated at 0.9 ppm, thus the L22 methyl resonance at 0.51 ppm may escape saturation.

A close examination of the crystal structure reveals that L22 D1 methyl may experience a high-field shift from the proximal aromatic rings and thus it may be the one to escape saturation in the STD-NMR experiment. Therefore, we repeated the optimization by excluding L22 D1 saturation. We simultaneously optimized six parameters and two torsion angles, τ_1 (C4–C5–C7–C′) and τ_2 (C5–C7–C1′–C2′) of TMP ligand. This optimization yielded a very low NOE R-factor of 0.055 with the following optimized parameters: $\tau_L = 0.082$ ns; $\tau_P = 20.60$ ns; $\tau_{m1} = 0.50$ ps; $\tau_{m2} = 3.96$ ps; $\tau_{m3} = 4.16$ ps; the leakage factor = 0.065 s^{-1}; $\tau_1 = -80°$ and $\tau_2 = 100°$. The STD buildup curve for all the five protons is shown in Fig. 12. The optimized protein correlation time is reasonable for the size of DHFR [80]. The new torsion angles ($\tau_1 = -80°$ and $\tau_2 = 100°$) are very close to the crystal structure values ($\tau_1 = -85°$ and $\tau_2 = 110°$). We note that this SICO structure was strictly based on STD intensities as experimental constraints. In principle it can be further improved by energy minimization using CHARMM or AMBER.

The lower steady state STD value of C7 methylene protons despite their very close proximity to the protein surface is a consequence of strong dipolar relax-

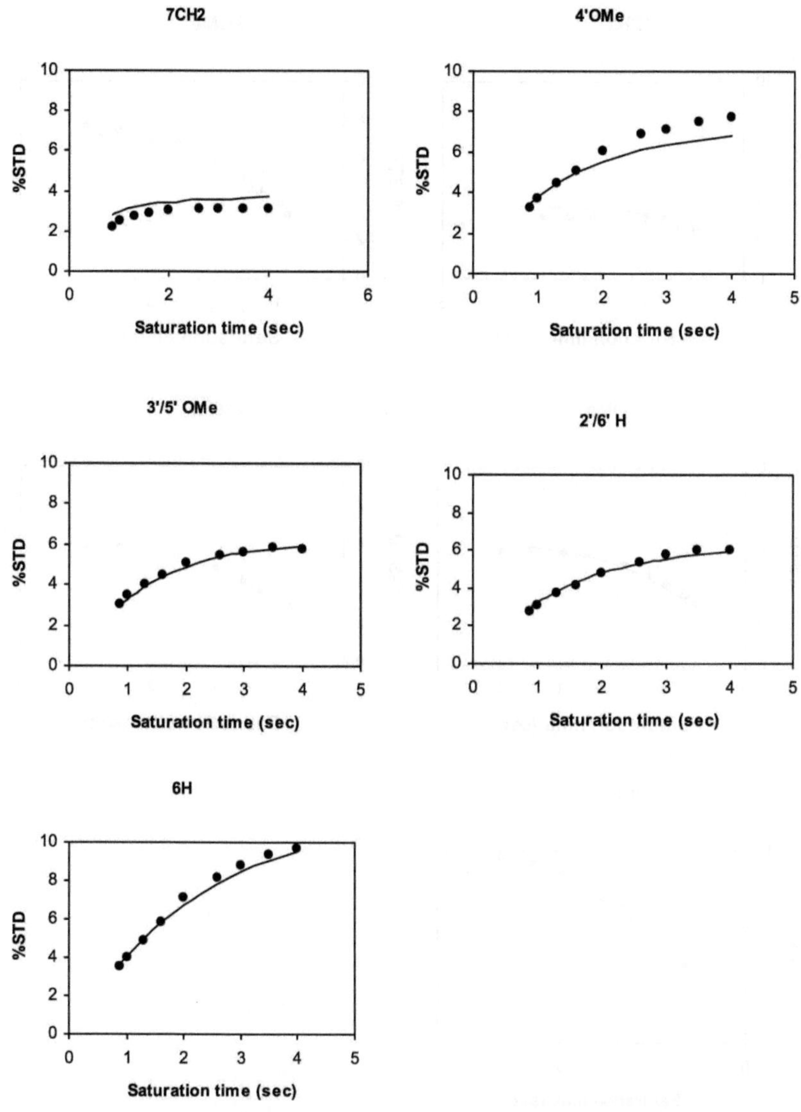

Fig. 11 Comparison of experimental STDs (*shaded circles*) and calculated STD values (*solid line*) from CORCEMA-ST method for the crystal structure of chicken liver DHFR/TMP complex. The optimized parameters are: $\tau_L = 0.101$ ns; $\tau_P = 20.43$ ns; $\tau_{m1} = 0.81$ ps; $\tau_{m2} = 3.04$ ps; $\tau_{m3} = 3.26$ ps; leakage factor = 0.065 s^{-1}; and NOE R-factor = 0.076. Reprinted with permission from [75]; © 2005, American Chemical Society

ation between the geminal protons. The significance of the differences in the relaxation rates of ligand protons on the STD values has been pointed out by us previously. In particular we showed that the geminal protons on a ligand are likely to show smaller STDs despite their close proximity to the protein [36].

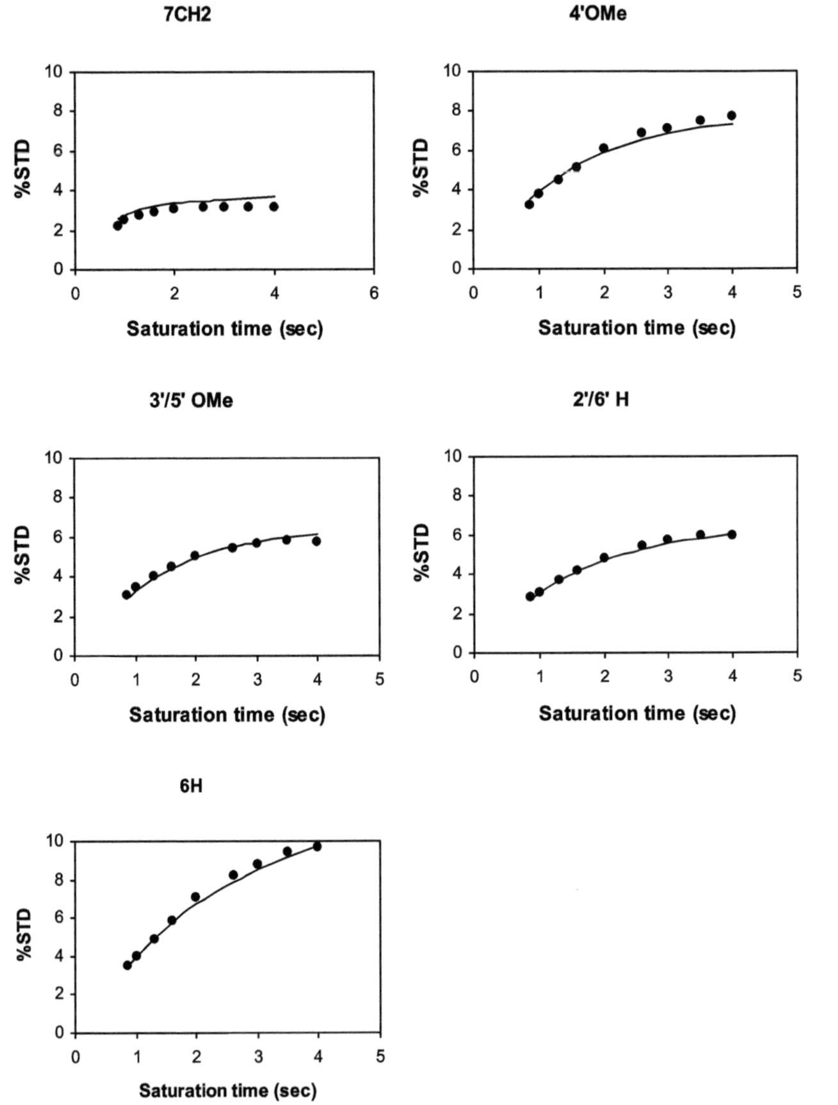

Fig. 12 Comparison of experimental STDs (*shaded circles*) and calculated STD values (*solid line*) from CORCEMA-ST method for the modified crystal structure ($\tau_1 = -80°$ and $\tau 2 = 100°$) of chicken liver DHFR/TMP complex. The optimized parameters are: τ_L = 0.082 ns; τ_P = 20.60 ns; τ_{m1} = 0.5 ps; τ_{m2} = 3.96 ps; τ_{m3} = 4.16 ps; leakage factor = 0.064 s^{-1}; and NOE R-factor = 0.055. Reprinted with permission from [75]; © 2005, American Chemical Society

The CORCEMA-ST calculations on the initial bovine DHFR/TMP model already yielded a very low NOE R-factor of 0.076, and this was further improved (NOE R-factor = 0.055) with a very slight refinement of two torsion angles (τ_1

and τ_2 of TMP ligand changed by 5° and 10°, respectively). This low R-factor indicates an excellent agreement between the experimental and CORCEMA-ST predicted STD values (Fig. 12), and reflects the high-quality experimental STD data set in [77], as well as the absence of any resonance overlap. This exceptionally low R-factor also serves as a validation of the CORCEMA-ST methodology developed in our laboratory.

The final CORCEMA-ST optimized structure (with a NOE R-factor of 0.055) of bovine DHFR/TMP complex is shown in Fig. 13. The CORCEMA-ST calculations of STD intensities show that the solution structure of bovine DHFR/TMP is essentially identical with the crystal structure of highly homologous chicken liver DHFR/TMP complex [76]. The torsion angles we obtained for the bovine DHFR/TMP complex ($\tau_1 = -80°$ and $\tau_2 = 100°$) compare favorably with those found in the crystallographic structures of the TMP/NADPH ternary complexes of chicken liver DHFR (–85°, 102°) studied by Matthews et al. [76] and mouse L1210 DHFR (–96°, 85P°) studied by Stammers et al. [81]. Most recently, the NMR structure of the ternary complex of human DHFR was reported [82] in which the τ_1 and $\tau 2$ values for TMP were found to be –153P° (or 207°) and 82°, in reasonable agreement with the values for the mouse DHFR-TMP-NADPH structure of Groom et al. [83],

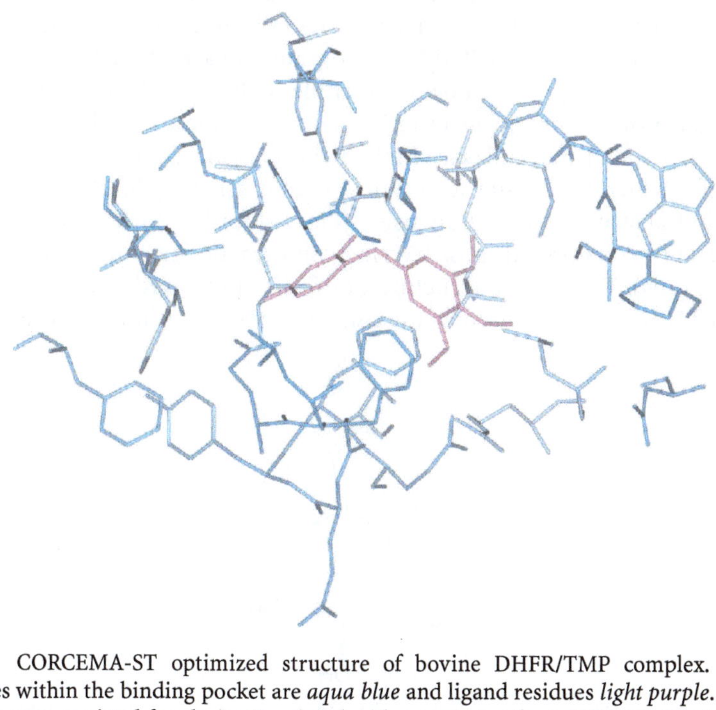

Fig. 13 CORCEMA-ST optimized structure of bovine DHFR/TMP complex. Protein residues within the binding pocket are *aqua blue* and ligand residues *light purple*. The hydrogens were omitted for clarity. Reprinted with permission from [75]; © 2005, American Chemical Society

but differing from those in the chicken liver, mouse L1210, and bovine DHFR complexes. The reason for this difference is not clear. Since the PDB coordinates for all these structures are now available, they are worth investigating by a combined STD-NMR and CORCEMA-ST analyses.

Our study underscores the significance of a quantitative interpretation of STD intensities using the CORCEMA-ST procedure for the characterization of binding epitopes at atomic resolution. In contrast to what was sometimes suggested [84], we show here that the STDs involving methyl resonances also give accurate information about epitopes in quantitative analyses.

6.4 STD-NMR/Molecular Modeling/CORCEMA-ST Protocol for Predicting the Reliable Binding Mode of Protein–Ligand Complexes in Aqueous Solution

In the absence of a detailed crystallographic or NMR structure of a protein–ligand complex, alternative methods have been developed that provide structural information on macromolecular complexes [85–87]. Recently, a combined STD-NMR/molecular modeling/CORCEMA-ST protocol has been developed for identifying the proper binding modes of ligands within protein binding pockets [88]. In this protocol the various binding modes generated by Autodock were evaluated by comparing the experimental STD intensities to those predicted by the CORCEMA-ST program for each of the binding modes to select the binding mode most compatible with the NMR data. The validity of this protocol has been tested with the complexes of the glycosidase inhibitors kifunensine and salacinol to the enzyme Golgi-mannosidase II (GMII) for which the crystal structure is already known [89, 90].

The initial structures of kifunensine (Fig. 14) for the docking studies were constructed by making all possible ring conformations using Sybyl 6.6 (Tripos, St. Louis). The crystal structure of dGMII complexed with kifunensine (PDB ID#1PS3) [89] was used as the control model for evaluating the success

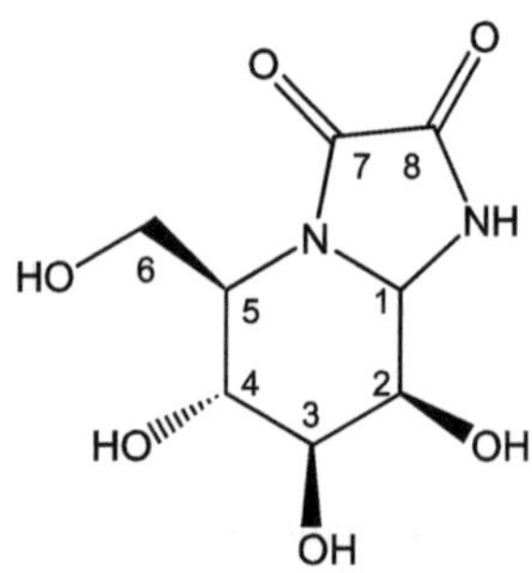

Fig. 14 Structure of kifunensine. Reproduced with permission from [90], © 2005, American Chemical Society

of this protocol in correctly identifying the proper binding mode. Automated docking was performed with Autodock 3.0 [91]. The details of the docking procedure can be seen in [88]. For each starting structure of kifunensine, 100 docked structures were obtained by using the Lamarckian genetic algorithm (LGA) searches and clustered according to the results differing in positional root-mean-square deviation (rmsd). The analysis of the docking results indicated that six conformers could be complexed with dGMII in three different binding modes (Fig. 15), based on the distance between the zinc ion and the hydroxyl groups of kifunensine. A key feature of binding mode I is that the distance between the zinc ion and the 2- or 3-hydroxyl groups of kifunensine is less than 2.3Δ. In mode II the closest distance between the polar group of kifunensine and the zinc ion is 3.31Δ. Mode III displayed a very different binding mode from I and II. Previous studies of complexes of dGMII inhibitors with different binding affinities have suggested the importance of coordination with the zinc ion to give a high-affinity inhibitor [89, 91].

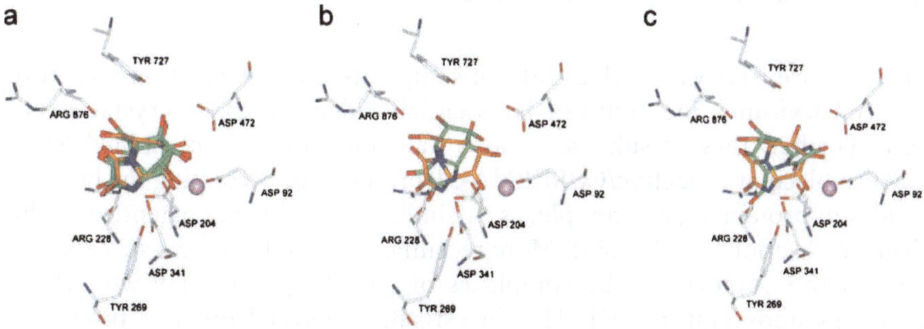

Fig. 15 Three binding modes of kifunensine in the active site of dGMII from docking calculations: **a** mode I **b** mode II and **c** mode III. The zinc ion is shown as a *pink ball*, and the relevant side chains are shown as *sticks*. Atoms in kifunensine are shown in different colours (*red* oxygen, *blue* nitrogen, and *green* carbon). Each binding mode of kifunensine to dGMII (*green* carbon atoms) is superimposed on the ligand in the corresponding crystal structure (*orange* carbon atoms) in the active site of dGMII. Reproduced with permission from [90], © 2005, American Chemical Society

The mode of binding predicted by the modeling study (mode I) was confirmed by STD-NMR experiments performed on the complex. In these studies, 20–30% STD enhancements were observed for seven unexchangable protons in kifunensine, as shown in Fig. 2 of [88]. Theoretical STDs were predicted by using the CORCEMA-ST procedure for all the three binding modes obtained from molecular modeling. As shown in Fig. 16, STD values predicted for mode 1 is in excellent agreement with the experimental values and yielded a very low R-factor of 0.24. However, mode II and mode III exhibited different patterns, as shown in Fig. 16. Thus mode II and III could be excluded. The mode I STD values are also in agreement with the crys-

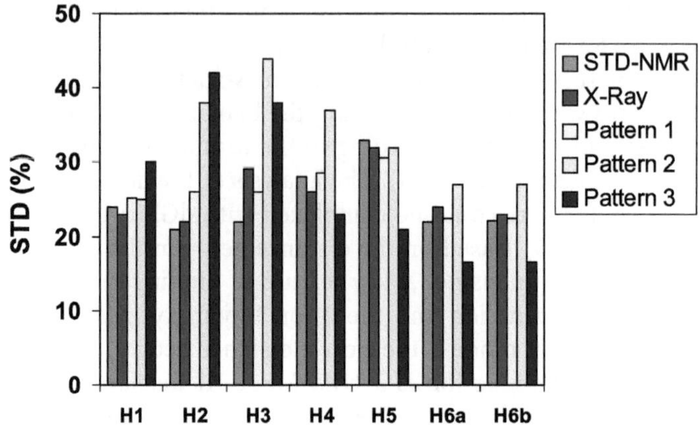

Fig. 16 Comparison of experimental and predicted STD values from the CORCEMA-ST protocol kifunensine-dGMII complex. *Pattern 1*, *pattern 2* and *pattern 3* refer to STD intensities calculated for modes I, II and III, respectively. Reproduced with permission from [90], © 2005, American Chemical Society

tal structure STD values (Fig. 16). This suggests that the solution structure of dGMII/kifunensine complex is essentially identical to its crystal structure. Further, these results serve as a validation of their combined STD-NMR/molecular modeling/CORCEMA-ST protocol for predicting the binding modes of protein–ligand complexes. A similar approach was extended to the dGMII/salacinol complex [88]. More recently, these authors have successfully applied this protocol to the complexes of UDP-Galp and UDP with UDP-galactopyranosemutase [93]. This investigation allowed them to probe the mechanism of interaction of UDP-Galp with the enzyme even though there is no crystallographic structure available for the complex.

7
Conclusions

In this review, we have summarized the CORCEMA-ST methodology along with some experimental examples where this method has been utilized successfully in obtaining quantitative structural information on the conformations of ligands within the binding pockets of large proteins. We have developed two versions of the CORCEMA-ST program. The first version is useful to simply predict the expected STD intensities for any given model of a ligand–protein complex for a given set of parameters such as correlation times, spectrometer frequency, and binding constant. Thus, this version can be used to verify if a crystallographic structure of a complex or a model based on a related complex is compatible with the experimental STD data. In the absence of such crystallographic structures, this version can be used to examine if any of the binding modes predicted by computer docking programs are compatible with solution NMR data [88, 93]. The second version is useful in refining the torsion angles of the bound ligand within the protein binding pocket of a protein to identify the global minimum conformation.

Even though in the two examples we presented in this work [69, 70, 75] we had utilized crystallographic structures as the starting structures before refinement, we wish to emphasize that these structures were used only to properly orient the ligand within the binding pocket, and not to have a bound ligand conformation close to the global minimum. This was because we have demonstrated that our second version of the program (with SICO protocol) can successfully locate the global minimum conformation for the ligand even when the starting structure is widely different. Thus, the results of these SICO refinements are most meaningful if the ligand is oriented properly in the binding pocket. Often however, such information may not be readily available due to lack of crystallographic data. In such instances, as demonstrated in recent elegant studies [88, 93], one can utilize robust computer docking programs (such as AutoDock or FlexiDock) to generate plausible binding modes, and test each one using a combined STD-NMR/molecular modeling/CORCEMA-ST protocol to identify the optimum model as judged by the lowest NOE R-factor. If necessary, the torsion angles of this model can be further refined using the SICO protocol to generate a better model with even lower NOE R-factors. The results of SICO optimization are strictly based on NMR constraints only, and can benefit by an additional step of energy minimization using CHARMM or AMBER force fields to further optimize the energy of interaction. If a group on the bound ligand is located outside the binding pocket, it may not be likely to experience significant STDs; thus its conformation may not be well defined in the SICO optimizations.

In the examples mentioned here there was the implicit assumption that the bound ligand does not alter the conformation of the protein, and that that the structure of the free protein can be used for the bound state. In those instances where the bound ligand causes a major conformational change in the binding pocket region (induced fit binding), the applicability of the CORCEMA-ST method may be limited in the absence of PDB coordinates for the modified binding pocket of the protein. The CORCEMA-ST theory presented here is applicable only to the case where there is only one type of ligand in the NMR sample that is weakly binding to a target protein (Scheme 1) with a 1 : 1 stoichiometry. If the NMR sample solution has two or more ligands that are competing to occupy the same binding pocket of a protein, or if the stoichiometry of the complex deviates from 1 : 1, the CORCEMA equations need to be modified to take into account the complicated equilibrium between several ligands with different affinities and several ligand–protein complexes in solution.

This method is primarily meant for weak complexes (with dissociation constants K_d in the range of 10^{-3} to 10^{-7} M) and utilizes STD NMR intensities as experimental constraints. Since many lead compounds bind target proteins weakly with affinities in the millimolar to micromolar range [19], the CORCEMA-ST procedure is likely to be of value in determining the bound conformations of these lead compounds positioned within the binding pocket of the target protein. This methodology is a powerful tool for analyzing the STD data and it puts STD-based epitope mapping on a quantitative basis. The combined use of STD-NMR data and CORCEMA-ST calculations is likely to be useful in structure-based drug design efforts.

Acknowledgements This work was supported in part by the NCI grant CA-13148. The authors wish to thank Prof. B. Mario Pinto for supplying the originals of some figures used in this review. The work on sialoadhesin/sialyllactose and UDP-Gal/galctosyltransferase complexes was from a fruitful collaboration with Prof. Thomas Peters at the Medical University of Lübeck.

References

1. Meyer B, Weimar T, Peters T (1997) Eur J Biochem 246:705
2. Chen A, Shapiro MJ (1998) J Am Chem Soc 120:10258
3. Chen A, Shapiro MJ (2000) J Am Chem Soc 122:414
4. Klein J, Meinecke R, Mayer M, Mayer B (1999) J Am Chem Soc 121:5336
5. Mayer M, Meyer B (1999) Angew Chem Int Ed Engl 35:1784
6. Vogtherr M, Peters T (2000) J Am Chem Soc 122:6093
7. Dalvit C, Pevarello P, Tato M, Veronesi M, Vulpetti A, Sundstorm M (2000) J Biomol NMR 18:65
8. Dalvit C, Fogliatto G, Stewart A, Veronesi M, Stockman B (2001) J Biomol NMR 21:349
9. Jahnke W, Perez LB, Paris CG, Strauss A, Fendrich G, Nalin CM (2000) J Am Chem Soc 122:7394
10. Hajduk PJ, Olejniczak ET, Fesik SW (1997) J Am Chem Soc 119:12257
11. Yan J, Kline AD, Mo H, Zartler ER, Shapiro MJ (2002) J Am Chem Soc 124:9984
12. Moore JM (1999) Biopolymers 51:221
13. Stockman BJ, Farley KA, Angwin DT (2001) Methods Enzymol 338:230
14. McCoy MA, Wyss DF (2000) J Biomol NMR 18:189
15. McCoy MA, Wyss DF (2002) J Am Chem Soc 124:11758
16. Medek A, Hajduk PJ, Mack J, Fesik SW (2000) J Am Chem Soc 122:1241
17. Shuker SB, Hajduk PJ, Meadows RP, Fesik SW (1996) Science 274:1531
18. Wyss DF, McCoy MA, Senior MM (2002) Curr Opin Drug Discov Devel 5:630
19. Meyer B, Peters T (2003) Angew Chem Int Ed Engl 42:864
20. Peters T, Biet T, Herfurth L (2003) Biol Magn Reson 20:287
21. Peng JW, Moore J, Abdul-Manan N (2004) Prog Nucl Magn Spectrosc 44:225
22. Lepre CA, Moore JM, Peng JW (2004) Chem Rev 104:3641
23. Mayer M, Meyer B (2001) J Am Chem Soc 123:6108
24. Haselhorst T, Weimar T, Peters T (2001) J Am Chem Soc 123:10705
25. Johnson MA, Pinto BM (2002) J Am Chem Soc 124:15368
26. Maaheimo H, Kosma P, Brade L, Brade H, Peters T (2000) Biochemistry 39:12778
27. Mayer M, James TL (2002) J Am Chem Soc 124:13376
28. Benie JB, Moser R, Bauml E, Blass D, Peters T (2003) J Am Chem Soc 125:14

29. Claasen B, Axmann M, Meinecke R, Meyer B (2005) J Am Chem Soc 127:916
30. Mari S, Serrano-Gomez D, Canada FJ, Corbi AL, Jimenez-Barbero J (2004) Angew Chem Int Ed Engl 44:296
31. Hajduk PJ, Mack JC, Olejniczak ET, Park C, Dandliker PJ, Beutal BA (2004) J Am Chem Soc 126:2390
32. Deng H, Cahill S, Kurz L, Callender R (2004) J Am Chem Soc 126:1952
33. Moseley HNB, Curto EV, Krishna NR (1995) J Magn Reson Ser B 108:243
34. Moseley HNB, Lee W, Arrowsmith CH, Krishna NR (1997) Biochemistry 36:5293
35. Krishna NR, Moseley HNB (1999) Structure Computation and Dynamics in Protein NMR. Biol Magn Reson 17:223
36. Jayalakshmi V, Krishna NR (2002) J Magn Reson 155:106
37. Krishna NR, Jayalakshmi V (2002) J Korean Magn Reson 6:94
38. Takahashi H, Nakanish T, Kami K, Arata Y, Shimada I (2000) Nat Struct Biol 7:220
39. Nakanishi T, Miyazawa M, Sakakura M, Terasawa H, Takahashi H, Shimada I (2002) J Mol Biol 318(2):245–249
40. Ramos A, Kelly G, Hollingworth D, Pastore A, Frenkiel T (2000) J Am Chem Soc 122:11311
41. Lipari G, Szabo A (1982) J Am Chem Soc 104:4546
42. Dellwo MJ, Wand AJ (1993) J Am Chem Soc 115:1886
43. Baleja JD, Pon RT, Sykes BD (1990) Biochemistry 29:4828
44. Brüschweiler R, Case DA (1994) Prog NMR Spect 26:27
45. Krishna NR, Agresti DG, Glickson JD, Walter R (1978) Biophys J 24:791
46. Xu Y, Sugar IP, Krishna NR (1995) J Biomol NMR 5:37
47. Torrey HC (1949) Phys Rev 76:1059
48. Wagner G, Wuthrich K (1979) J Magn Reson 33:675
49. Jayalakshmi V, Krishna NR (2004) J Magn Reson 168:36
50. Zabell AP, Post CB (2002) Proteins 46:295
51. Meadows RP, Hajduk PJ (1995) J Biomol NMR 5:41
52. Fahmy A, Wagner G (2002) J Am Chem Soc 124:1241
53. Metropolis N, Rosenbluth AW, Rosenbluth MN, Teller AH, Teller E (1953) J Chem Phys 21:1087
54. Xu Y, Krishna NR (1995) J Magn Reson Ser B 108:192
55. Alotto PG, Eranda C, Brandstatter B, Furntratt G, Magele G, Nervi M, Preis K, Repetto M, Richter KR (1998) IEEE Trans Magn 34:3674
56. Ng KK, Weis WI (1997) Biochemistry 36:979
57. Driscoll PC, Gronenborn AM, Beress L, Clore GM (1989) Biochemistry 28:2188
58. Frank M, Lang AB, Wetter T, Lieth CW (2002) In Silico Biol 2:38
59. Borgias BA, James TL (1989) Methods Enzymol 176:169
60. Mertz JE, Guntert P, Wuthrich W, Braun W (1991) J Biomol NMR 1:257
61. Bonvin AM, Boelens R, Kaptein R (1991) J Biomol NMR 1:305
62. Borgias BA, James TL (1988) J Magn Reson 79:493
63. Bhunia A, Jayalakshmi V, Benie AJ, Schuster O, Kelm S, Krishna NR, Peters T (2004) Carbohydr Res 339:259
64. May AP, Robinson RC, Vinson M, Crocker PR, Jones EY (1998) Mol Cell 1:719
65. Huyer W, Neumaier A (1999) J Global Optimization 14:331
66. Crocker PR, Vinson M, Kelm S, Drickamer K (1999) Biochem J 341:355
67. Palmer AG, Case DA (1992) J Am Chem Soc 114:9059
68. Ramakrishnan B, Balaji PV, Qasba PK (2002) J Mol Biol 318:491
69. Jayalakshmi V, Biet T, Peters T, Krishna NR (2004) J Am Chem Soc 126:8610
70. Jayalakshmi V, Biet T, Peters T, Krishna NR (2005) J Am Chem Soc 127:7261

71. Sugawara BY, Iwasaki H (1984) Acta Cryst C 40:389
72. Gastinel LN, Cambillau C, Bourne Y (1999) EMBO J 18:3546
73. Petrova P, Koca J, Imberty A (2001) Eur J Biochem 268:5365
74. Biet T, Peters T (2001) Angew Chem Int Ed 40:4189
75. Jayalakshmi V, Krishna NR (2005) J Am Chem Soc 127:14080
76. Mathews DA, Bolin JT, Burridge JM, Filman DJ, Volz KW, Kaufman BT, Beddell CR, Champness JN, Stammers DK, Kraut J (1985) J Biol Chem 260:381
77. Yan J, Allen DK, Mo H, Shapiro MJ, Zartler ER (2003) J Magn Reson 163:270
78. Feeney J (2000) Angew Chem Int Ed Engl 39:291
79. Stockman BJ, Nirmala NR, Wagner G, Delcamp TJ, De Yarman MT, Freisheim JH (1992) Biochemistry 31:218
80. Searle MS, Forster MJ, Birdsall B, Roberts GCK, Feeney J, Cheung HTA, Kompis I, Geddes AJ (1988) Proc Natl Acad Sci USA 85:3787
81. Stammers DK, Champness JN, Beddell CR, Dann JG, Eliopoulos E, Geddes AJ, Ogg D, North ACT (1987) FEBS Lett 218:178
82. Kovalevskaya NV, Smurnyy YD, Polshakov VI, Birdsall B, Bradbury AF, Frenkiel T, Feeney J (2005) J Biomol NMR 33:69
83. Groom CR, Thillet J, North ACT, Pictet R, Geddes AJ (1991) J Biol Chem 266:19890
84. Berteau O, Sandstorm C, Bielicki J, Anson DS, Kenne L (2003) J Am Chem Soc 125:15296
85. Kelley MD, Mancera RL (2004) J Chem Inf Comput Sci 44:1942
86. Klebe G (2000) J Mol Med 78:269
87. Johnson MA, Pinto BM (2004) Carbohydr Res 339:907
88. Wen X, Yuan Y, Kuntz DA, Rose DR, Pinto BM (2005) Biochemistry 44:6729
89. Shah N, Kuntz DA, Rose DR (2003) Biochemistry 42:13812
90. Kuntz DA, Ghavami A, Johnson BD, Pinto BM, Rose DR (2005) Tetrahedron:Asymmetry 16:25
91. Morris GM, Goodsell DS, Halliday RS, Huey R, Hart WE, Belew RK, Olson AJ (1998) J Comput Chem 19:1639. See also http://autodock.scripps.edu/, last visited: 8 August 2007
92. van den Elsen JM, Kuntz DA, Rose DR (2001) EMBO J 20:3008
93. Yuan Y, Wen X, Sanders DAR, Pinto BM (2005) Biochemistry 44:14080

Top Curr Chem (2008) 273: 55–116
DOI 10.1007/128_2007_18

Published online: 4 April 2008

Structural and Functional Studies of Peptide–Carbohydrate Mimicry

Margaret A. Johnson[1] (✉) · B. Mario Pinto[2]

[1]Department of Molecular Biology, The Scripps Research Institute, 10550 North Torrey Pines Rd., MB-44, La Jolla, CA 92037, USA
maggie@scripps.edu

[2]Department of Chemistry, Simon Fraser University, Burnaby, B.C., V5A 1S6, Canada

Abstract Certain peptides act as molecular mimics of carbohydrates in that they are specifically recognized by carbohydrate-binding proteins. Peptides that bind to anti-carbohydrate antibodies, carbohydrate-processing enzymes, and lectins have been identified. These peptides are potentially useful as vaccines and therapeutics; for example, immunologically functional peptide molecular mimics (mimotopes) can strengthen or modify immune responses induced by carbohydrate antigens. However, peptides that bind specifically to carbohydrate-binding proteins may not necessarily show the corresponding biological activity, and further selection based on biochemical studies is always required. The degree of structural mimicry required to generate the desired biological activity is therefore an interesting question. This review will discuss recent structural studies of peptide–carbohydrate mimicry employing NMR spectroscopy, X-ray crystallography, and molecular modeling, as well as relevant biochemical data. These studies provide insights into the basis of mimicry at the molecular level. Comparisons with other carbohydrate-mimetic compounds, namely proteins and glycopeptides, will be drawn. Finally, implications for the design of new therapeutic compounds will also be presented.

Keywords Molecular modeling · NMR spectroscopy · Peptide–carbohydrate mimicry · Protein–ligand interactions · X-ray crystallography

1 Introduction

1.1 Peptide–Carbohydrate Mimicry—General Aspects

The use of carbohydrates as drugs and vaccines has several limitations. Molecular mimics of carbohydrates provide an alternative source of compounds to target pathways involving protein–carbohydrate interactions. Molecular mimicry is gaining importance as a way of finding therapeutic compounds with the desired biological activity of natural carbohydrates, but with potential advantages such as greater specificity or selectivity, or lower toxicity. In recent years, immunological studies have demonstrated the ability of certain peptides to act as molecular mimics of carbohydrates, in that they are able to induce anti-carbohydrate immune responses. Carbohydrate-mimetic peptides that bind to enzymes and lectins have also been discovered. The nature of this mimicry at the molecular level is currently the subject of investigation.

Intuitively, the observation of biologically functional mimicry might imply structural mimicry: it would seem likely that compounds producing similar biological effects by interaction with a receptor would have similar shapes or similar arrangements of functional groups. Indeed, some examples of molecular mimicry in immunology involve peptides derived from microbial proteins, which share fortuitous sequence similarity with self proteins, giving rise to harmful cross-reactions against self tissues and autoimmune disease [1, 2]. This situation is conceptually simple to rationalize, as peptides with similar sequences might have similar conformational preferences, even if presented in the context of different proteins. Structural mimicry is also easily rationalized for the glycosidase inhibitors of the alkaloid class of natural products [3, 4], as these have some structural similarity to carbohydrates, and at least some similar interactions with the protein receptor are likely to be responsible for molecular mimicry.

In contrast, the existence of peptide–carbohydrate mimicry is more surprising in that it is difficult to picture how these two different classes of compounds could "mimic" each other. The origin of this effect at the molecular level has been the subject of recent investigations by NMR spectroscopy, X-ray crystallography, and molecular modeling. In combination with functional data, these studies provide insight into the nature of this phenomenon.

1.2 Peptide–Carbohydrate Mimicry in Immunology

Carbohydrates are an important class of molecules in the treatment of infectious disease. Many bacteria produce carbohydrates as part of the cell wall (Gram-positive bacteria) or cell membrane (Gram-negative bacteria). Furthermore, many species of bacteria secrete exopolysaccharide capsules which surround the cell and act as physical barriers, and also possess defensive properties against the human immune system; the capsule is often essential for virulence [5, 6]. These carbohydrates are often the first molecules recognized by the human immune system and would in theory be good vaccine candidates. In addition to their presence on the surfaces of microbial pathogens, carbohydrates coat the surfaces of mammalian cells, and altered glycosylation occurs frequently in cancer. Certain oligosaccharides are characteristically expressed on cancerous cells and represent excellent targets for anticancer vaccines or drugs [7, 8].

The use of carbohydrates as vaccines poses several challenges: Since carbohydrates are not usually processed by antigen-presenting cells or displayed by MHC molecules, they do not stimulate T cells or lead to a memory response, class switching and affinity maturation. The antibody response is usually limited to IgM and is of weak affinity. A successful strategy has been to conjugate polysaccharides to protein carriers, which often leads to a greatly improved response, and has resulted in several successful vaccines [5, 6]. An alternative strategy would be to use mimotopes—peptides which act as immunologically functional molecular mimics of carbohydrates—either to replace the carbohydrate molecule as a vaccine, or to supplement it in order to boost the immune response. Another possibility offered by peptide mimotopes is that of selecting one desirable component of the immune response, and down-regulating other undesirable components. Given these possibilities, it is perhaps not surprising that the majority of the currently known examples of peptide–carbohydrate mimicry have resulted from immunological research and involve interactions with antibodies.

2 Background and Context

2.1 Background

The first reports of carbohydrate-mimetic peptides appeared in 1992 when Oldenburg et al. [10] and Scott et al. [11] identified peptides that bound to the lectin concanavalin A, for which the natural ligands are α-linked mannopyranosides. Employing a randomized library of hexapeptide sequences displayed on the surface of filamentous phage, Scott et al. identified the peptide MYWYPY. This peptide was shown to bind specifically to concanavalin A, and its binding was inhibited by methyl α-D-mannopyranoside. In a separate report, Oldenburg et al. identified different peptides also containing the motif YPY, which were also recognized specifically by concanavalin A [10]. In 1993, Hoess et al. [12] reported the first carbohydrate-mimetic peptide specific for an antibody receptor, the octapeptide APWLYGPA. The antibody was directed against the Lewis Y (Le^Y) tumor antigen, and the octapeptide is a possible candidate in the development of cancer vaccines.

These initial reports showed that functional carbohydrate-mimetic peptides could be found. These peptides acted as ligands for several biologically important proteins, and had potential in the development of therapeutics. Subsequently, peptides that act as molecular mimics of carbohydrates in binding to a wide variety of different receptors have been identified [14–44]: these are summarized in Table 1 and reviewed in [13]. Some of these peptides were able not only to bind to a particular receptor, but also to act as immunologically functional mimics and to induce an anti-carbohydrate immune re-

Table 1 Carbohydrate-mimetic peptides

Publication date	Peptide[1]	Protein receptor	Carbohydrate target	Refs.
1994	SFGSGFGGGY	anti-Glc*N*Ac mAbs; Glc*N*Ac-specific lectins	Glc*N*Ac	[15, 16]
1995	HFVQH	mIgA C5	*Shigella flexneri* 5a LPS	[17]
1995	DITWDQLWDLMK RNMSWLELWEHMK	E-selectin	sialyl Lewis X	[18]
1995	cyclo-iWyRyN	yeast α-glucosidase	α-linked glucosides	[19]
1996	GLQYTPSWMLVG SYSWMYE	mAb 2H1	*Cryptococcus neoformans* capsular polysaccharide (glucuronoxylomannan)	[20–22]
1997	YPPYFTLMY RNVPPIFNDVYWIAF	mAb AD117m and mAb H11; *Ricinus communis* lectin; β-galactosidase	lactotetraosylceramide, neolactotetraosylceramide	[23]
1997	YDWLMF WLWEW	mAb 12A1, mAb 13F1	*Cryptococcus neoformans* capsular polysaccharide (glucuronoxylomannan)	[24]
1997	DRPVPY	mAb SA-3	Group A *Streptococcus* cell-wall polysaccharide	[25]
	MCPPLYSPSACA	mAb Strep 9	Group A *Streptococcus* cell-wall polysaccharide	
	ADAAPSPTPYLPRLS	mAb HGAC 39	Group A *Streptococcus* cell-wall polysaccharide	
	MCRPGIPTHHCA	mAb HGAC 47	Group A *Streptococcus* cell-wall polysaccharide	
	SCISAACFCI	mAb HGAC 101	Group A *Streptococcus* cell-wall polysaccharide	
	MDWNMH	mAb SYA/J6	*Shigella flexneri* Y LPS	
1998	DAHWESWL	IB4 lectin; anti-α-Gal Abs	Gal-α-(1-3)-Gal	[26, 27]
1998	WHWRHRIPLQLAAGR	mAb KA17	GD1α ganglioside	[28]
1998	yryygl	mAb HGAC 39	Group A *Streptococcus* cell-wall polysaccharide	[29]
1998	WENWMMGNA	mAb S9	Group B *Streptococcus* capsular polysaccharide	[30]
1999	PHCKVNRGC	mAb SEAM 28	*N. meningitidis* B polysaccharide (poly-α-(2-8)-sialic acid)	[31]
1999	PPSFCPNFIPCTDGL, YGPWCDDTPQACRSWP	mAb A15-6B3, mAb B66-2C8	*Brucella abortus* and *Brucella melitensis* LPS (poly-α-4-formamido-perosamine)	[32]

Table 1 (continued)

Publication date	Peptide[1]	Protein receptor	Carbohydrate target	Refs.
1999	KLWQLPV, TFGLQSL	mAbs 10, 227, 109, L3 58	ganglioside GA1 (asialo GM1)	[196]
2000	DLWDWVVGKPAG	mAb NS19-9	sialyl Lewis A	[33]
2000	IELLQAR	mAb 7LE; E-selectin	Lewis A; sialyl Lewis X	[34]
2001	DHQRFFV	mAb HMenB1	*N. meningitidis* B polysaccharide (poly-α-(2-8)-sialic acid)	[35]
	SHVPNAF	mAb HMenB3 mAb	*N. meningitidis* B polysaccharide (poly-α-(2-8)-sialic acid)	
	VVSTGSH	HMenB13	*N. meningitidis* B polysaccharide (poly-α-(2-8)-sialic acid)	
2001	YDSeYeSe, EGEWPVYP	RHAMM	hyaluronan (poly-(β-(1-4)-GlcA-β-(1-3) -GlcNAc)-)	[36]
2002	NIKFNPP	mAb Hyp6BM1	*S. pneumoniae* type 6B capsular polysaccharide	[37]
	TTPGPWF	mAb Hyp6BM8		
2002	FLHTRLFVSDWYHTP	mAb L2-412	GlcA-5-sulfate-β(1-3)-Galβ(1-4)GlcNAc	[38]
2002	YEW	tetanus toxin	GT1b ganglioside	[39]
2004	DYlYF TYDYF	galectins	galactose, lactose	[40]
2004	CGAVIDDC CRGDDFPAWYC CAQQWPEWYPC	mAb 9-2-L3,7,9	*N. meningitidis* type L3,7,9 lipopolysaccharide (lactotetraose)	[41]
2004	cyclo-CSSVTAWTTGCG cyclo-CSLIASMETGCG	anti-PSA mAb	poly-α-(2-8)-sialic acid	[42]
2005	LFSPWLRVQNHF VAAVTEKLRAGT	mAb MG96	unknown epitope on mosquito midgut glycoproteins; contains Man-α-(1-6)- and LacNAc components	[43]
2005	CGRLKMVPDLEC CDGGWLSKGSWC	mAb ch14.18	ganglioside GD2 (Galβ(1-4)-(NeuAc-α-(2-8)-NeuAc-α-(2-3)-Galβ-(1-4)Glcβ(1-1)-Cer)	[44]

[1] Many studies have isolated several groups of peptides; in these cases, only one or a few representative peptide sequences are given.

sponse. The first example of such an immunogenic mimic was reported by Westerink et al. [45], who constructed a 13-residue peptide based on the CDR H3 sequence of an anti-idiotypic antibody. This antibody mimics the *Neisseria meningitidis* C capsular polysaccharide, and is protective against meningococcal infection in animals. The synthetic peptide was also shown to induce a protective anti-polysaccharide response [45, 46].

The first example of an immunogenic mimic discovered by phage-display library screening was reported by Phalipon et al. [47], who isolated 19 carbohydrate-mimetic peptides from nonapeptide libraries by screening with monoclonal antibodies (mAbs) directed against the *Shigella flexneri* type 5a lipopolysaccharide O-antigen. Two of these sequences were able to induce an immune response against the lipopolysaccharide and against this bacterial strain. Immunogenic peptide mimics have been referred to as mimotopes, by analogy to the immunological term epitope, meaning the specific topography or structure within a molecule, that is recognized by an antibody. More recent studies demonstrated the ability of mimotopes to induce immune responses against the polysaccharides of several bacteria (group B *Streptococcus*, *N. meningitidis*, *N. gonorrheae*, *Streptococcus pneumoniae*, *Haemophilus influenzae*), against cell-surface carbohydrates of fungi and parasites (*Cryptococcus neoformans*, *Candida albicans*, *Entamoeba histolytica*); and against tumor-associated carbohydrate antigens and carbohydrate antigens of the HIV virus [45–78]. These reports, summarized in Table 2, illustrate that functional peptide–carbohydrate mimicry exists and demonstrate the great potential of peptides in vaccine development.

While functional (immunological) mimicry has been established, the basis of mimicry on the molecular level remains to be explained. Several hypotheses have been put forward; one of the earliest was that the side chains of aromatic amino acid residues might mimic the hydrophobic faces of the pyranosyl rings of carbohydrates. Before 1997, no structural evidence was available to support or discount these hypotheses. The nature of peptide–carbohydrate mimicry on the molecular level became the subject of structural investigations, and the resulting studies along with functional data will be discussed below.

2.2 Definition of Mimicry

With the goal of exploring the specificity of peptide–carbohydrate mimicry, Harris et al. [25] undertook an investigation of peptide–carbohydrate mimicry using a panel of related anti-carbohydrate antibodies rather than a single receptor. Phage-displayed libraries were used to identify peptide ligands for several closely related antibodies all directed against the same carbohydrate, a branched trisaccharide representing the minimal epitope of the cell-wall polysaccharide of group A *Streptococcus* [79]. Carbohydrate-mimetic

Table 2 Mimotopes, or carbohydrate-mimetic peptides capable of inducing anti-carbohydrate immune responses

Publication date	Peptide	Protein receptor	Carbohydrate target	Refs.
1995	CARIYYRYDGTAY CARIYYRYDGFAY	mAb 1E4	*N. meningitidis* C polysaccharide (poly-α-(2-9)-sialic acid)	[45, 46, 48]
1997	CYKPLGALTHC CKVPPWARTAC	mIgA I3 mIgA I3, C5	*Shigella flexneri* 5a lipopolysaccharide	[47]
1997	YYPYD, YYRYD, YYRGD	antibodies against tumor-associated Lewis antigens	tumour-associated Lewis antigens	[49, 50]
1997	DVFYPYPYASGS	concanavalin A; anti-α-methyl-mannoside Abs	α-methyl-mannosides	[51-53]
1998	APWLYGPA	mAb B3	Lewis Y	[54]
1998	FDTGAFDPDWPA	mAb S9	Group B *Streptococcus* capsular polysaccharide	[30]
2000	FSLLW	mAb BR55	Lewis Y	[55, 56]
2000	GEASGLCCRWSSLKGC	mAb 9C10	*N. meningitidis* A polysaccharide	[57]
2000	SMYGSYN, APARQLP	mAb 9-2-L3,7,9	*N. meningitidis* type L3,7,9 lipopolysaccharide (lactotetraose)	[58]
2000	ASRNKANDYTTEY SASVKGRFIVS	anti-*S. pneumoniae* Abs	*S. pneumoniae* cell wall polysaccharide	[59]
2001	GMDGTQLDRW	mAb 2E9	*C. neoformans* polysaccharide (glucuronoxylomannan)	[60, 61]
2001	SGQARVLYSEFINAL	mAb4	*Streptococcus pneumoniae* 4 capsular polysaccharide	[62]
2001	FHENWPS	mAb DJ2.8	*Candida albicans* phosphopeptidomannan (β-(1-2)-linked mannosides)	[63]
2001	RYGRY	concanavalin A	α-linked mannosides displayed on HIV envelope protein	[51, 64]
2001		mAb 2C7	*N. gonorrheae* lipooligosaccharide	[65]
2002	GTHPXL	mAb EH5	*Entamoeba histolytica* proteophosphoglycans	[66, 67]

Table 2 (continued)

Publication date	Peptide	Protein receptor	Carbohydrate target	Refs.
2002	RDEMQASMWS HKWVMLWNYG CGQSHIADC CHLWRNYDC	mAbs M4.37.3, PC1.4.1, PC2	*S. pneumoniae* cell wall polysaccharide	[68]
2002	cyclo-ACNTIGGYECGGGSK cyclo-ACSWLHMPYCGGGSK	mAb 9-2-L3,7,9	*N. meningitidis* type L3,7,9 lipopolysaccharide (lactotetraose)	[69]
2003	NMMRFTSQPPNN NMMNYIMDPRTH	polyclonal anti-*H. influenzae* Abs	*H. influenzae* lipooligosaccharide	[70]
2004	GFSYYRPPWIL	mAb 1E4	*N. meningitidis* C polysaccharide (poly-α-(2-9)-sialic acid)	[71]
2004	NKVIWDRDWMYP NKVIWEADWAFS HHSHSILQSDWF	mAb HmenB3	*N. meningitidis* group B polysaccharide (poly-α-(2-8)-sialic acid)	[72]
2005	FHLPYNHNWFAL	mAb NAD (human IgA)	*S. pneumoniae* type 8 capsular polysaccharide	[73]
2005	KVWFLPEAAQPS, EARVFSSKHWIP, TSSALKCCFIQ	mAb 2G12; concanavalin A	α-linked mannosides displayed on HIV envelope protein	[74]
2005	QEPLMGTVPIRAGGGS	polyclonal Abs against *M. tuberculosis* neutral polysaccharides	*M. tuberculosis* arabinomannan/ lipoarabinomannan	[75]
2005	EDPSHSLGLDVALFM	mAb 14G2a	ganglioside GD2 (Galβ (1-4)-(NeuAc-α-(2-8)-NeuAc-α-(2-3)-Galβ(1-4)Glcβ(1-1)-Cer)	[76]
2005	IPVLDENGLFAP	mAb 2C7	*N. gonorrheae* lipooligosaccharide	[77]
2005	AEGEFYCSGPP DRVCWGPDPAK; AEGEFCSPPD SPGVCGDPAK	mAb Vc1	*Vibro cholerae* O139 capsular polysaccharide	[78]
2006	DRPVPY	mAb SA-3	Group A *Streptococcus* cell-wall polysaccharide	[25, 138]

peptides were found, but the peptides selected differed between each antibody, and the peptide that bound to each antibody was not recognized by the others. Thus, mimicry depended on the receptor considered. Therefore, the peptides could not be considered general mimics of the branched trisaccharide, but rather mimics of its binding to one specific protein. In general, peptide and carbohydrate ligands should not be considered molecular mimics of each other "by definition", and furthermore, carbohydrate-mimetic peptides do not have general properties that allow them to mimic a particular carbohydrate (as can be seen from an inspection of Tables 1 and 2, the peptide sequences selected are essentially random and do not share identifiable chemical characteristics). Rather, mimicry is with respect to interactions with a receptor, the same interactions which confer binding affinity.

2.2.1
Structural Mimicry

The observation that the mimetic peptide is specific for the receptor—not specific for the carbohydrate it mimics—implies possible differences in the nature of ligand/receptor interactions. Nonetheless, to provide strong binding affinity for a receptor, mimetic ligands might in part share some structural features with the natural ligand (structural mimicry). As a framework for discussion, we will define structural mimicry as occurring when a mimetic ligand makes similar contacts to the receptor binding site as the natural ligand. This implies that in its bound conformation, the mimetic ligand will provide certain functional groups in spatially equivalent positions as the natural ligand. These functional groups should be chemically similar to those of the natural ligand, although not necessarily identical (for example, hydrogen bond donors and acceptors; or hydrophobic groups). Some hypothetical examples of structural mimicry are shown in Scheme 1.

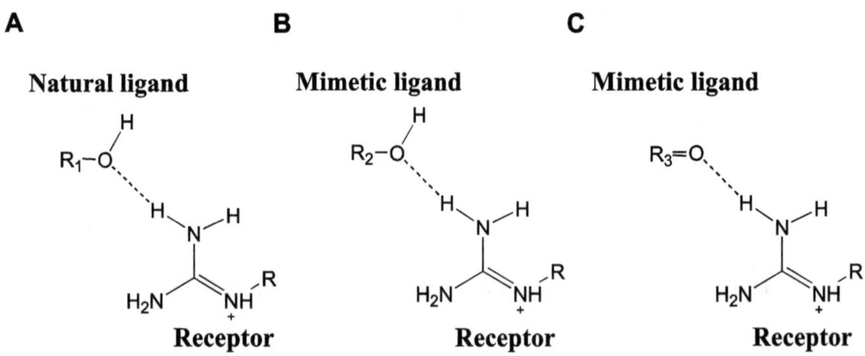

Scheme 1 Examples of structural mimicry

Alternatively, binding affinity between the mimetic ligand and receptor might be achieved by a different mechanism than that of the natural ligand. The mimetic ligand might contact a different area of the receptor binding site, or provide a different set of intermolecular interactions. As shown in

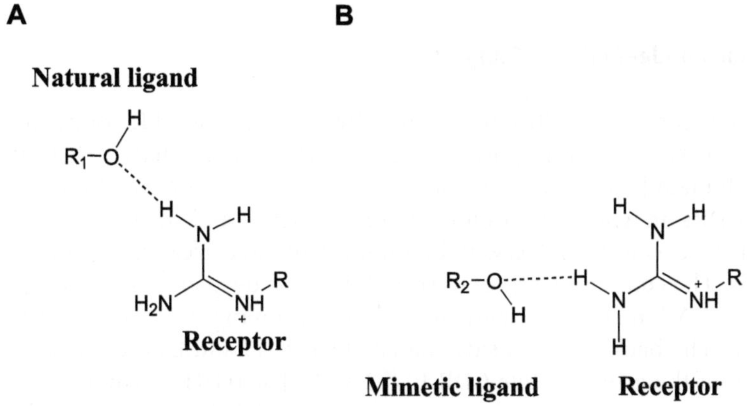

Scheme 2 Alternatives to structural mimicry

Scheme 2, in this case the mimetic ligand and natural ligand might be found in different positions within the binding site of the receptor.

3
The Origin of Mimicry in an Antibody Combining Site

The crystal structures of the complexes between an anti-carbohydrate antibody and its peptide [80] and oligosaccharide [81, 82] ligands, allowed for the first time in 2002 a direct examination of the nature of peptide–carbohydrate mimicry. The monoclonal antibody (mAb) SYA/J6 is a murine IgG_3 antibody directed against the polysaccharide O-antigen of the lipopolysaccharide of *Shigella flexneri* type Y [83]. The polysaccharide is a linear heteropolymer of L-rhamnopyranose (6-deoxy-L-mannopyranose; Rha*p*) and 2-acetamido-2-deoxy-D-glucopyranose (Glc*p*NAc) residues [84]:

$$[\rightarrow 2)\text{-L-Rha}p\text{-}\alpha\text{-}(1\rightarrow 2)\text{-L-Rha}p\text{-}\alpha\text{-}(1\rightarrow 3)\text{-L-Rha}p\text{-}\alpha\text{-}(1\rightarrow 3)\text{-D-Glc}pN\text{Ac-}\beta\text{-}(1\rightarrow]$$

A B C D

The structures of the antibody Fab fragment in complex with the octapeptide MDWNMHAA, discovered by screening of phage-displayed libraries [25], and with a pentasaccharide, (ABCDA′, or (L-Rha-α-(1-2)-L-Rha-α-(1-3)-L-Rha-α-(1-3)-D-Glc*N*Ac-β-(1-2)-L-Rha-α-OMe), representing a segment of the O-antigen, were solved by X-ray crystallography. This allowed a detailed comparison of specific intermolecular interactions in the complexes, and revealed the basis of peptide–carbohydrate mimicry in this system [80].

3.1 Pentasaccharide–Antibody Complex

Antibody combining sites are formed by six hypervariable loops, or complementarity-determining regions (CDRs), which link adjacent β-strands in the V domain [85, 86]. The light and heavy chains associate and each donates three CDRs, in which variation in length, sequence and structure is responsible for the existence of Abs with binding affinity to a great variety of different ligands. The combining site of mAb SYA/J6 has a deep groove shape, approximately 25 Å long by 12 Å wide by 10 Å deep, sloping to a deep pocket in the bottom. The backbone and side chains of CDR H3 and L3 are exposed at the bottom of the groove, while CDR L1/L2 and CDR H1/H2 form the two sides. These features are designed to accommodate a left-handed helical shape of the polysaccharide (Fig. 1C).

The pentasaccharide ABCDA′ fits into the groove with the C rhamnosyl residue buried most deeply in the site, where it makes multiple van der Waals interactions with Met H100A, Thr L91 and Gly H99 at the bottom of the groove. Less deeply buried are the B rhamnosyl and D GlcNAc residues. The Rha B 3- and 4-OH groups form hydrogen bonds to the Ala H97 carbonyl group and to the Tyr L32 side chain hydroxyl group, respectively. The

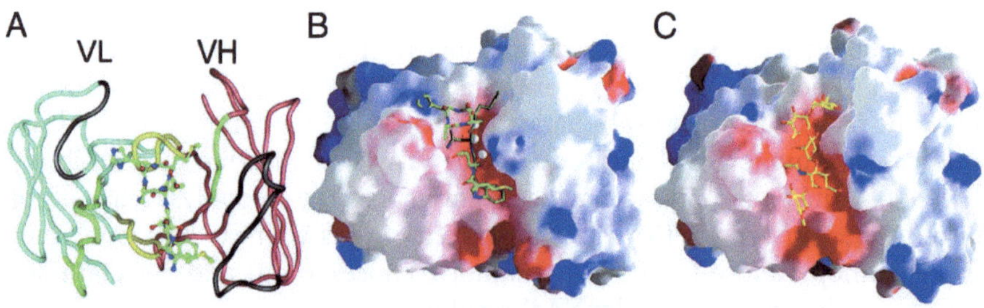

Fig. 1 Views of SYA/J6 antibody complexes with octapeptide and pentasaccharide ligands. **A** The antibody–peptide complex, with the peptide shown as a ball-and-stick model and the antibody as a backbone trace. The variable regions of the light (V_L) and heavy (V_H) chains are colored *blue* and *mauve*, respectively. The CDRs of both the light and heavy chains are shown in *green* (L1 and H1); *gray* (L2 and H2); and *yellow* (L3 and H3). **B** The antibody–peptide complex with the antibody shown as a molecular surface, colored according to electrostatic potential (−10 kT, *red*; neutral, *white*; + 10 kT, *blue*), and with the peptide represented as a stick model. In both (**A**) and (**B**), the atom types for the peptide are C, *green*; N, *blue*; and O, *red*. Also shown are the three ordered water molecules located in the deepest part of the combining site groove (*white spheres*; from top, S2, S9, and S1). (**C**) The antibody–pentasaccharide complex, similar view as in (**B**), with the sugar residues in the order Rha A, Rha B, Rha C, GlcNAcD, and Rha A′, from *top* to *bottom* (data from [82]). The atom types for the sugar are C, *yellow*; N, *blue*; and O, *red*. Rha C mainly occupies the deepest part of the groove (groove pocket). Reproduced from [80].

GlcNAc D *N*-acetamido group is positioned with the methyl group in a hydrophobic pocket (formed by Tyr L32, His L27D, and Rha C) while the NH group donates a hydrogen bond to Thr L91 O; its 4-OH group hydrogen bonds to the side chain of Glu H50. The *N*-acetamido group and the 4-OH are very important for binding, as shown by greatly decreased binding affinity of deoxygenated analogs [87, 88]. In the pentasaccharide complex, the A and A′ Rha residues are the most exposed and contact residues at the edges of the combining site, e.g. His L27D and Trp H33. These residues still contribute to the binding affinity, as the pentasaccharide binds more strongly than the trisaccharide BCD by 0.6 kcal/mol [82].

3.2 Peptide–Antibody Complex

Like the pentasaccharide, the octapeptide lies along the groove, roughly parallel to the V_H–V_L interface (Fig. 1A,B). One observes immediately that the peptide contacts some of the same areas of the binding site as the saccharide, but also extends into other areas of the site. The first four residues, Met P1-Asp P2-Trp P3-Asn P4, adopt an extended conformation, and the last four residues, Met P5-His P6-Ala P7-Ala P8, form one turn of α-helix.

The *N*-terminal end of the peptide is placed in the end of the groove where the reducing end of the sugar is normally found ("bottom" in the view of Fig. 1). Here, the Met P1-Asp P2 side chains are somewhat exposed to solvent and interact weakly with the binding site. The following three side chains, Trp P3-Asn P4-Met P5, lie along the bottom of the groove, while the α-helical turn of residues His P6-Ala P8 allows these side chains to interact with residues at the top of the binding site.

Residues Met P1 and Asp P2, in the same area as Rha A′, interact with the antibody only by two hydrogen bonds and very few van der Waals interactions (Fig. 3A,C). Structural mimicry occurs here in that Asp P2 OD1 acts as a hydrogen bond acceptor to His L27D NE2 in a similar way as Rha A′ O3. Additional interactions, not representative of structural mimicry, are also present: Met P1 contacts H58, a residue not contacted by the pentasaccharide. Asp P2 forms water-mediated hydrogen bonds with the Fab: a water molecule (S5) bridges Asp P2 N and the NH2 guanidino nitrogen of Arg H52, while Asp P2 O hydrogen bonds directly to the other (NH1) guanidino nitrogen (Figs. 2B, 3A). In the pentasaccharide complex, a water molecule is coordinated by Glc*N*Ac D O1 and O5, and is located 3.62 Å from NH1 of the Arg H52 guanidino group. In this case, Asp P2 O is close to the position of the water molecule in the pentasaccharide complex, although in the latter complex, the distance between the water molecule and Arg H52 is too long to represent a hydrogen bond, and is not considered to represent structural mimicry. The pentasaccharide complex has no interaction corresponding to the Asp P2 N–H_2O–Arg H52 NH2 interaction, as the pentasaccharide does not have spatially equivalent atoms near the NH2 guanidino nitrogen.

The residues Trp P3–Asn P4–Met P5 lie in the center of the groove, roughly coincident with the sugar residues Rha C and GlcNAc D. Like the two sugar residues, this part of the peptide is the most deeply buried and forms the greatest number of interactions. A striking feature of the peptide complex is the involvement of a large number of ordered water molecules, and interestingly, one of these could be considered to provide structural mimicry of a polar group of the sugar. Thus, the ordered water molecule S6 is found in a similar position as the amide N of GlcNAc D, and forms hydrogen bonds to the backbone amide nitrogen of Trp P3 and to Thr L91 O and Thr L92 O. The interaction with Thr L91 O could be considered structural mimicry, as it is similar to that of the GlcNAc D 2-NH group. The peptide provides a second hydrogen bond to Thr L91 O from Asn P4 ND2.

The indole ring of Trp P3 lies in a hydrophobic pocket formed by residues from CDR L3 (Val L94, Pro L95); CDR H2 (His H58) and a framework residue (Trp H47, FR2) (Fig. 4A). As shown in Fig. 1, the saccharide does not enter this pocket. The side chain of Met P5 lies in close contact with the indole ring of Trp H33 (Fig. 4B); the pentasaccharide forms one hydrogen bond with this residue, but no hydrophobic interactions.

The peptide does not directly contact the central, deepest part of the groove, where Rha C binds. Rather, three water molecules occupy this cavity (Figs. 1B, 2B). These are hydrogen bonded to each other in the order S2-S9-

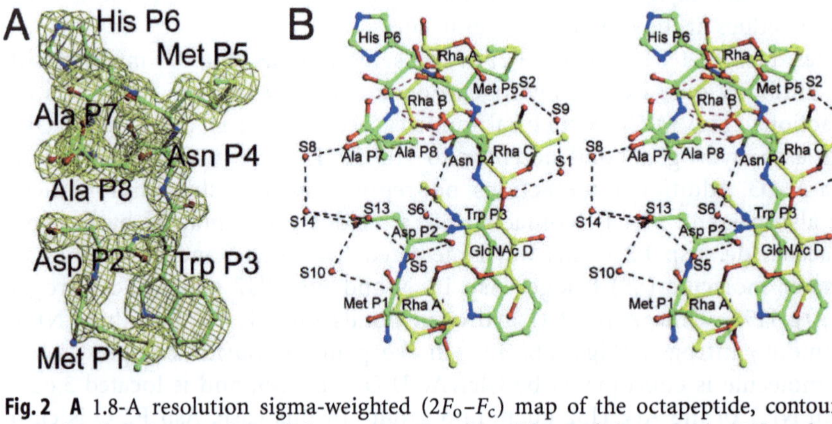

Fig. 2 **A** 1.8-A resolution sigma-weighted ($2F_O$–F_C) map of the octapeptide, contoured at 1σ level. **B** Stereo view of the superimposed positions of the peptide [80] and carbohydrate [82] ligands. Atom types are the same as in Fig. 1. The ordered solvent (water) molecules (S) associated with the peptide complex are also shown. The solvent molecules S2, S9 and S1 and the residue Rha C are shown to occupy a similar area of the site, the deep pocket of the combining site groove (see also Fig. 1). Reproduced from [80]. © 2003 by The National Academy of Sciences of the USA

S1. S2 and S1 form hydrogen-bonded bridges between the peptide and Fab. S9 is found at the deepest point in the site, in a similar position to the Rha C 6-methyl group. Trp P3 O coordinates water S1, which hydrogen bonds to Glu H35, an interaction which is similar, though not identical, to the sugar's essential interaction with the nearby Glu H50 (GlcNAc D O4–Glu H50). Water S2 is coordinated by Met P5 N and hydrogen bonds to Gly H96 O; this is similar to a polar interaction observed in the pentasaccharide complex (Rha C O4–Gly H96 O, 3.68 Å) although this distance is too long to represent a hydrogen bond, and the corresponding interaction does not represent structural mimicry. Overall, the three water molecules S2-S9-S1 serve both as intermolecular bridges and to provide a topography that is complementary to the combining site.

The α-helical turn of residues Met P5–Ala P8 exposes the last four peptide side chains for specific interactions, most of which differ from those in the pentasaccharide complex. The only structurally mimetic hydrogen bond here is from the side chain of His P6 to Tyr L32 OH, which is similar to the hydrogen bond from Rha B 4-OH in the pentasaccharide complex. An additional interaction of the peptide is another hydrogen bond of the His P6 side chain

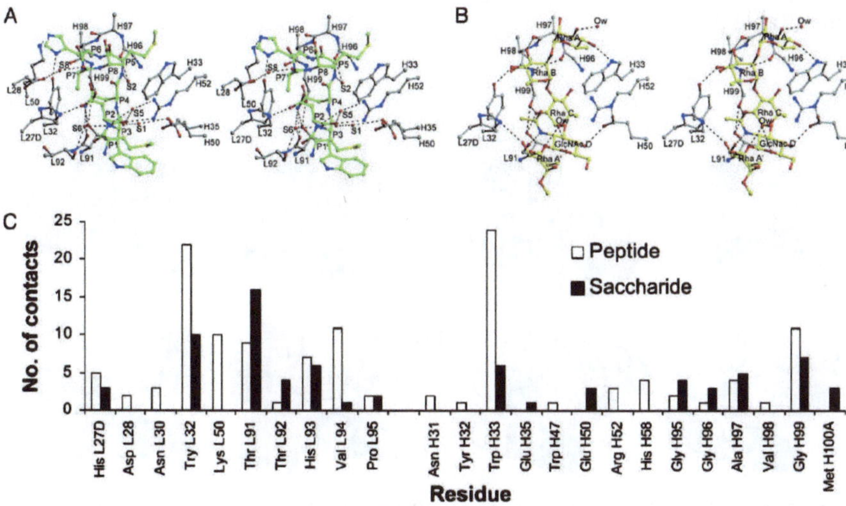

Fig. 3 Interactions between the Fab and octapeptide or pentasaccharide. **A** Stereoview of octapeptide binding, which includes eight direct and nine one-water-mediated hydrogen bonds. **B** Stereoview of pentasaccharide binding, adapted from [82]. The two water molecules (Ow) bind solely to the sugar. **C** Comparison of the number of direct contacts of residues between the Fab and octapeptide or pentasaccharide. The numbers of direct hydrogen bonds and van der Waals contacts ($\leq$4 Å) associated with peptide binding are 6 and 126, respectively, and with saccharide binding are 8 and 74, respectively. Reproduced from [80]. © 2003 by The National Academy of Sciences of the USA

to Lys L50 NZ, the first residue of the short loop CDR L2—an unusual interaction because CDR L2 is usually not contacted in complexes of antibodies with small molecules, peptides and carbohydrates [89–92], and is not contacted by the pentasaccharide. Met P5 lies along the face of the indole ring of Trp H33 (Fig. 4B). Ala P7 and Ala P8, at the end of the turn, are at the top of the site. The Ala P7 side chain does not make direct contact with the site and is partially exposed, but its carbonyl O hydrogen bonds to Asp L28 via a water molecule. The Ala P7 side chain lies close to, but not completely inside, a small hydrophobic pocket (Tyr L32, His L27D, Thr L91) which in the sugar complexes accommodates the Glc*N*Ac D *N*-acetyl methyl group; this can be considered structural mimicry although the peptide is not perfectly complementary to the pocket as is the carbohydrate (Fig. 4C,D). Finally, the Ala P8 side chain lies close to the edge of the Trp H33 indole ring.

Overall, mimicry in this system consists of a small number of structurally mimetic interactions, with a greater number of interactions that do not represent structural mimicry. Furthermore, the peptide complex exhibits interesting differences from the pentasaccharide complex, including a large number of ordered water molecules: 14, vs. 2 in the pentasaccharide complex. This difference is likely related to differences in the thermodynamics of binding: for the peptide, a high enthalpy of binding is offset by an unfavorable entropy of binding, so that the overall binding affinities of the two ligands are almost the same (Table 3).

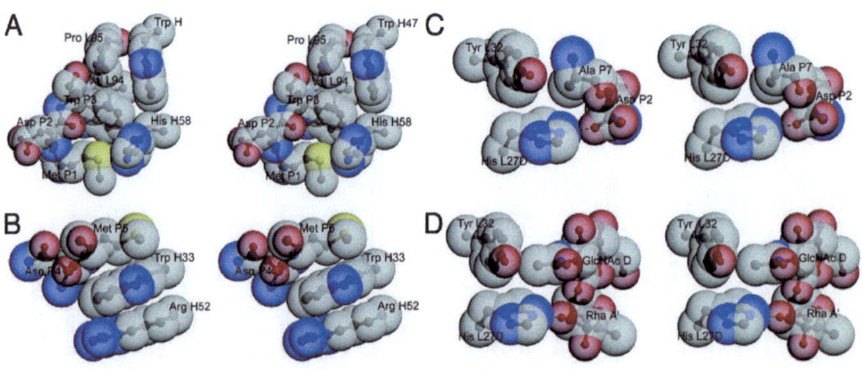

Fig. 4 Stereo views (in transparent, space-filling models) of selected hydrophobic interactions in the Fab-peptide and -pentasaccharide complexes. **A** Trp P3 buried in the hydrophobic cavity formed by CDR L3, His H58 and Trp H47. **B** Interactions of the side chain of Met P5 with Trp H33. **C** and **D** Interactions of the peptide and pentasaccharide, respectively, with His L27D and Tyr L32 (the pocket which accommodates the GlcNAc D methyl group). In (**C**) the hydrogen bond between the Asp P2 side chain and His L27D NE2 is also shown. A corresponding hydrogen bond between Rha A1′ 3-OH and His L27D NE2, shown in (**D**), represents an element of structural mimicry. Reproduced from [80]. © 2003 by The National Academy of Sciences of the USA

Table 3 Thermodynamics of binding of the peptide and pentasaccharide ligands with the antibody SYA/J6

	Peptide	Pentasaccharide
K_A (M^{-1})	5.7×10^5	2.5×10^5
ΔG (kcal mol^{-1})	- 7.9	- 7.4
ΔH (kcal mol^{-1})	- 16.9	- 1.5
$-T\Delta S$ (kcal mol^{-1})	9.0	- 5.9

The peptide is more highly complementary to the combining site, and makes a greater number of intermolecular contacts, 126 compared to the pentasaccharide's 74. A quantitative measure of complementarity is the Sc value, which is 0.77 for the peptide vs. 0.67 for the pentasaccharide (with 1.0 representing perfect complementarity). Interestingly, this value is increased to 0.82 if the three water molecules S2-S9-S1 are included, showing that in addition to providing intermolecular hydrogen bonds, these water molecules help to provide complementarity to the deepest part of the binding site.

If contacts to the different CDRs are considered, the overall pattern is similar. The pentasaccharide depends greatly on contacts to the H3 and L3 CDRs, at the bottom of the binding site; the peptide makes an almost equal number of contacts to this area of the site, despite having no direct contact with the deepest part of the groove, and in addition, exploits H1/H2 and L1/L2, located higher in the binding groove, to a greater extent. For both ligands, 57% of the contacts are made with the light chain, an unusual feature in complexes of antibodies with small-molecule ligands [92]; in such complexes, a greater number of interactions with the heavy chain is usually observed.

Structural mimicry of the carbohydrate by the peptide is observed for two direct hydrogen bonds to key residues in the site: Tyr L32 and His L27D, and for one water-mediated hydrogen bond to Thr L91; it is also observed for the hydrophobic interactions with the pocket of Tyr L32 and His L27D. These interactions are supplemented by a greater number of interactions of the peptide which are not structurally mimetic, including several other direct and water-mediated hydrogen bonds, and also the extensive hydrophobic interactions of Trp P3 and Met P5. Conversely, some important interactions of the sugar are not mimicked by the peptide. In the pentasaccharide complex, the Rha C 6-methyl group is completely buried at the bottom of the site, interacting with the side chain of Met H100A. The peptide does not contact this residue at all, and the area occupied by this side of the Rha C pyranosyl ring is occupied by the three water molecules S2-S9-S1. The differences in interaction patterns are summarized in Fig. 3C.

As described above, there are differences in the numbers of contacts of the two ligands to the antibody CDRs. There are 126 direct intermolecular contacts in the SYA/J6-peptide complex, and this number increases if water-mediated interactions are included. This is higher than in any other reported antibody–peptide complex, and is more comparable to the numbers of contacts in antibody–protein complexes, in which both the epitope and the Ab contact surfaces are flatter, leading to a large amount of buried surface area [93]. The contact with CDR L2 and the presence of bridging water molecules are also features which are more commonly observed in Ab–protein complexes [93–95]. These features suggest that the diversity of the phage-displayed library allowed the selection of a peptide with the maximum number of interactions with the site.

In summary, peptide binding includes some structural mimicry, comprising both hydrogen bonds and hydrophobic interactions. However, structurally mimetic interactions are in the minority, and binding appears to be mainly driven by a large number of other favorable interactions. Peptide binding involves a large number of ordered water molecules, which help to provide complementarity, especially at the deepest point of the binding site. Both light and heavy chains are contacted by the peptide, which effectively exploits a large area of the binding site.

3.3 Further Studies of the Octapeptide MDWNMHAA

3.3.1 Saturation Transfer Difference NMR Studies of the Peptide–Fab Complex

It is of interest to compare the structural results obtained from the crystal structure of the octapeptide–Fab complex with those of the complex in solution. In this case, transferred NOE enhancements were not observed, which is likely due to unfavorable binding kinetics. Saturation-transfer difference NMR (STD-NMR) [96, 97] employs intermolecular NOE transfer from a selectively saturated protein receptor to protons of a ligand that, when bound, are in close contact with receptor protons in the binding site (reviewed in [98, 99]; see also the work by N. Rama Krishna in this volume). Since a relatively long saturation period is employed, binding interactions may often be detected which do not give rise to transferred NOE effects. Furthermore, this technique allows epitope mapping, or the determination of which portions of a ligand are in direct contact with the protein receptor.

STD-NMR experiments applied to the MDWNMHAA/Fab complex allowed the identification of the peptide residues in direct contact with the antibody, and a comparison with the crystal structure [100]. Strong enhancements of all the ring protons of Trp P3 were observed, consistent with the extensive contacts of this residue with the antibody. In addition, the β-CH2 protons of Asn P4 were enhanced, and the methyl groups of Ala P7 and Ala P8. Enhancements were detected for both methionine ε-CH3 groups, with one much more strongly enhanced than the other. This is consistent with the observation that Met P5 makes extensive contact with Trp H33, while Met P1 is solvent-exposed and only contacts the edge of the binding site.

Some differences with respect to the crystal structure were also observed: for example, enhancements of the amide hydrogen atoms of His P6 and Met P5, and of the β-CH2 protons of His P6, were not observed, although they are in close proximity to multiple neighboring protons. It is possible that since the His P6 residue is located on the surface of the binding groove, flexibility might reduce the amount of contact with the antibody. The lack of enhancement of the Asn P4 side chain amide protons was also unexpected, as this side chain is buried deeply in the binding site. In general, the observed enhancements correlate well with the crystal structure (Fig. 5), but several inconsistencies indicate possible differences between the crystal and solution structures. More detailed analyses of the relaxation pathways involved in saturation [101], as well as the effects of flexibility and solvent-exchange rates, will likely be required to conclusively demonstrate the origin of these differences.

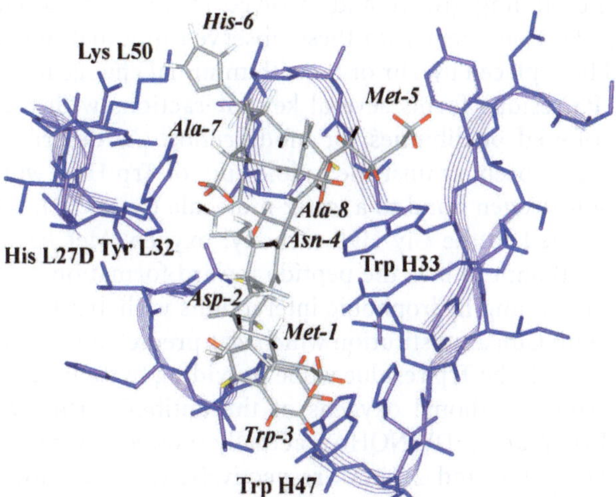

Fig. 5 The crystal structure of the antibody–octapeptide complex, with STD-NMR intensities mapped onto the bound peptide. Residues of the antibody combining site are shown in *purple*, with selected residues labeled, and the direction of the backbone indicated in ribbon representation. Residues of the peptide are labeled in italics. Heavy atoms of the peptide are shown in *gray*, while the default color for hydrogen atoms is *white*. Observed STD-NMR intensities are mapped onto hydrogen atoms of the peptide by color, with *red* indicating 50–100% enhancement, *orange* 30–50% enhancement, and *yellow* < 30% enhancement. Protons that are definitely *not* enhanced are shown in *black*; those for which no enhancement could be determined (due to interference by other resonances, or not observable in the 1D spectrum) remain *white*. Reproduced with permission from [100]. © 2004 Elsevier Science

3.3.2 Rational Design of Modified Peptides

Molecular modeling studies were undertaken with the aim of using the above-described structural data to design peptides with increased binding affinity to the antibody. Alanine-scanning studies [102] showed that substitution of Trp P3 or Asn P4 with alanine led to loss of binding, consistent with the structural observations of multiple interactions of these residues at the center of the combining site. Therefore, this area was not selected for optimization. However, the 1- and 5-positions of the peptide offered possibilities for modification to improve the binding affinity. Molecular modeling was performed with the structure of the octapeptide complex as a starting point for modification, followed by multiple rounds of manual docking and energy minimization to regularize the structures. Peptides with modifications that appeared favorable were synthesized, and their binding affinities were measured by competition ELISA assays, i.e. inhibition of antibody binding to polysaccharide O-antigen immobilized on a solid phase [103].

The Met P1 side chain is somewhat solvent-exposed, in contact only with the edge of the binding groove and involved in an intramolecular interaction with Trp P3. Consistent with these observed minimal interactions, this residue could be replaced by Thr or Ile with minimal change in binding affinity. The Met P5 residue forms several key interactions with the combining site but also offered possibilities for modification. As described above, its side chain stacks closely against the indole ring of Trp H33, while its amide group forms a hydrogen bond to a water molecule (water S2), which in turn is hydrogen-bonded to the Gly H96 carbonyl oxygen. Met P5 was replaced with Pro in an attempt to bias the peptide toward formation of the α-helical turn, while maintaining hydrophobic interactions with Trp H33. Met P5 was also replaced with Gln, a substitution which was predicted to allow hydrophobic interactions with the Trp residue while providing two additional hydrogen bonds to backbone carbonyl oxygens of the antibody. However, the peptides MDWNPHAA and MDWNQHAA actually showed lower binding affinity (IC_{50} values of 3.1 mM and 2.7 mM, respectively, vs. 19 μM for the original peptide). For the Pro substitution, this was possibly due to the loss of one water-mediated hydrogen bond, which could affect the water network at the bottom of the binding groove. Binding affinity could not be recovered by using, instead, 4-hydroxyproline at position 5. For the Gln replacement, the side chain's increased polarity may have adversely affected the hydrophobic interactions with the Trp ring.

In preparation for use as a surrogate vaccine, these peptides were synthesized as conjugates to tetanus toxoid, and in this form also competed with the O-antigen polysaccharide for binding [104]. Whether these peptides will function as effective immunogenic mimics will likely depend on their propensity to adopt the bioactive conformation in solution. A certain amount of structural mimicry may also be required for immunologically functional mimicry; alternatively, a large number of favorable contacts to both light and heavy chains, as observed in the octapeptide crystal structure, may be sufficient.

4 Mimicry of Other Bacterial Polysaccharides

4.1 Group B *Streptococcus*—Targeting a Conformational Epitope

Several peptide mimotopes have potential in the development of antibacterial vaccines. Group B streptococci are pathogenic bacteria which cause neonatal sepsis, meningitis, and pneumonia [105]. The capsular polysaccharide of type III GBS (GBSPIII) has a branched pentasaccharide repeating unit with terminal sialic acid residues. Progress has been made toward the development of a polysaccharide conjugate vaccine, but the polysaccharide is poorly immunogenic, and protective efficacy has not yet been satisfactorily demonstrated [30]. The poor immunogenicity may be due to a close resemblance to human carbohydrate structures that also contain terminal sialic acid residues (for example, NeuAc-α-(2-3)-Gal).

Antibodies produced against the polysaccharide recognize a length-dependent epitope, which has been characterized by ELISA and surface plasmon resonance studies using oligosaccharide fragments [106]. It was proposed that a certain conformation, only possible in long fragments with several repeating units, is recognized by these antibodies. Although the polysaccharide exists as a random coil in solution, populating a range of conformations, there may be transiently occurring secondary structures, such as helices, that present distinct molecular topographies [107]. Antibodies are likely to be directed at these conformational epitopes, in order to avoid recognition of the shorter sialic acid-containing oligosaccharides found on human tissues. Conformational epitopes such as these present unique opportunities for the use of peptide mimotopes as vaccines. If peptides that mimic the helical conformation of the polysaccharide could be identified, antibodies directed against this structure could perhaps be elicited more successfully.

Two peptide mimotopes were identified by screening of phage-display libraries [30]. The receptor employed was a monoclonal antibody (mAb S9) of the IgM class, directed against GBSPIII and known to be protective. The stronger binding peptide, with the sequence FDTGAFDPDWPA, was recognized not only by the isolating mAb but by two other mAbs and by polyclonal anti-GBS antibodies—an unusual case since most carbohydrate-mimetic peptides are recognized only by the isolating mAb. Immunization of mice with

Fig. 6 Views of the calculated bound structure of the peptide FDTGAFDPDWPA. **A** Backbone (N, Cα, C, O) atoms of the ten structures with lowest NOE restraint energy produced by simulated annealing refinement, superimposed using backbone atoms of residues 7–10. The lowest energy structure is colored by atom (C, *green*; N, *blue*; O, *red*), and a ribbon representation is shown in *green*. The other structures are shown in *light blue*. Residues 7–10 comprise a type I β-turn and are labeled. **B** Close-up views of residues 6–12 of the lowest energy calculated structure (colored by atom). Backbone atoms only are shown for residues 6 and 12. In the *left panel*, the turn is shown in an orientation similar to the structures in (**A**); in the *right panel*, it is shown in an orthogonal orientation, so that the exposed position of the Trp-10 side chain, on one face of the turn, is visible. **C** Close-up views of residues 6–12, as in (**B**), but in a typical type I β-turn conformation, as observed for the free peptide in solution. The conformation is very similar to the calculated bound conformation (**B**). Reproduced with permission from [108]. © 2003 The American Society for Biochemistry and Molecular Biology

the stronger-binding peptide conjugated to KLH elicited antibody responses against the purified polysaccharide and against whole GBS cells [30].

Solution NMR studies of the free peptide revealed the presence of a type I β-turn spanning the residues Asp 7 through Trp 10 [108]. The peptide existed as a mixture of *trans* and *cis* isoforms, in the ratio 7 : 3, due to isomerization about the Trp 10–Pro 11 peptide bond. The interaction of the peptide with the mAb S9 was investigated by NMR spectroscopy employing the transferred nuclear Overhauser effect [108]. The bound conformation was not well defined except in the region of Asp-7 to Pro-11, where a type I β-turn was observed (Fig. 6). Saturation transfer difference NMR (STD-NMR) experiments revealed enhancements of the Trp-10 H-7 resonance, as well as other aromatic resonances belonging either to Trp-10 and/or Phe-1 and Phe-6; and in addition, Thr 3, Ala 5, Asp-7, Pro-8, Pro-11 and Ala 12, indicating that these side chains are in contact with the antibody. The large number of STD enhancements in the Asp-7 to Pro-11 region is consistent with this region being a well-defined β-turn in the bound conformation. Such a turn would naturally expose these side chains for intermolecular interactions. This

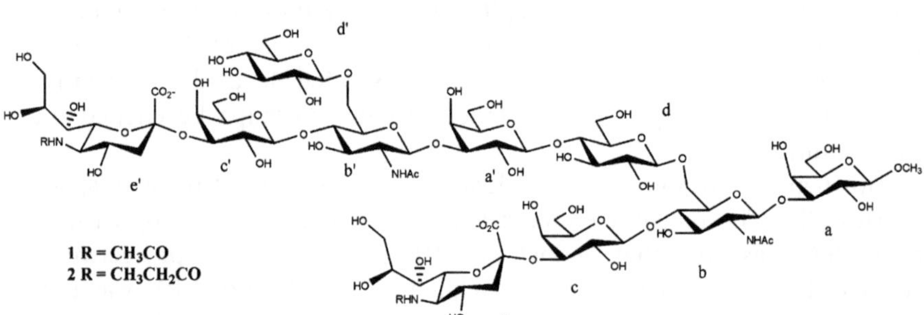

Scheme 3 Decasaccharides representing two repeating units of the GBS type III capsular polysaccharide (GBSPIII)

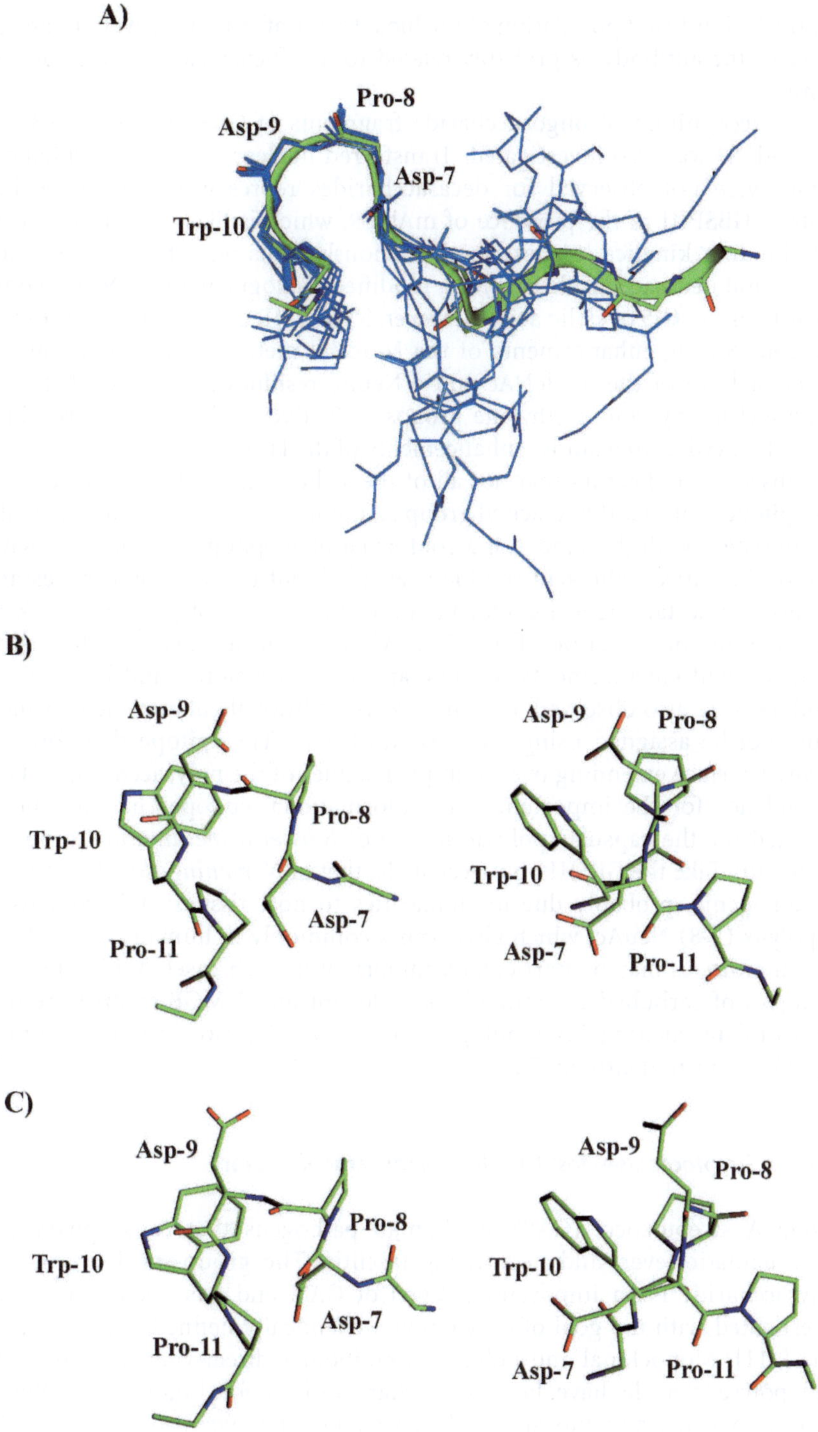
A)
Pro-8
Asp-9
Asp-7
Trp-10
B)
Asp-9
Pro-8
Trp-10
Asp-7
Pro-11
Asp-9
Pro-8
Trp-10
Asp-7
Pro-11
C)
Asp-9
Pro-8
Trp-10
Asp-7
Pro-11
Asp-9
Pro-8
Trp-10
Asp-7
Pro-11

peptide's significant population of the bioactive conformation, even in the absence of the antibody, is probably related to its efficacy as an immunogenic mimic.

The recognition of oligosaccharide fragments of GBSPIII (Scheme 3) by the mAb S9 was also investigated. Transferred nuclear Overhauser enhancements were not observed for decasaccharides representing two repeating units of GBSPIII in the presence of mAb S9, which is likely due to unfavorable binding kinetics, as described previously. This was the case both for the natural decasaccharide and for a modified analogue with an *N*-propionyl substituent on C-5 of sialic acid. However, STD-NMR enhancements were observable. Strong enhancements of the *N*-acetyl methyl groups of **1** (due to either or both of the b GlcNAc and e NeuAc residues), and also of the *N*-propionyl methyl and methylene groups of **2** (due to the e NeuAc residue) were observed. Importantly, enhancements of the H-3 protons of NeuAc were not observed, indicating that not all of the sialic acid residue contributes to the epitope, but that the *N*-acetyl group is important. This observation would be consistent with the model of a conformational epitope in which recognition of the entire sialic acid residue is avoided, but this residue is necessary to ensure a certain helical conformation of the entire polysaccharide, likely including the presentation of the *N*-acetyl group on the surface of the helix. Further small enhancements of the c and a (Gal) residues and b (GlcNAc) residues were also observed, and there were additional enhancements which could not be assigned using only 1D NMR data. The epitope therefore involves a surface extending over multiple residues of the polysaccharide [108].

Evidence for the importance of conformational epitopes has also been reported for the capsular polysaccharide of *Neisseria meningitidis* group B [109, 110]. Like the GBSIII polysaccharide, that of *N. meningitidis* B is poorly immunogenic, probably due to similarities to host tissues; it is composed of poly-α-(2-8)-NeuAc, which also occurs commonly in human tissues. Peptide mimotopes may be particularly important in such cases. While the first examples of carbohydrate-mimetic peptides for anti-NMGB mAbs were not successful in eliciting immunological mimicry [31], more recently, mimotopes have been identified [72].

4.2
Group A *Streptococcus*—Test Case for a Very Large Receptor

Group A streptococci (GAS) are human pathogens that cause pharyngitis, rheumatic fever, and necrotizing fasciitis. The group-specific cell-wall polysaccharide is an important antigen of GAS, and has been extensively investigated with the goal of developing a chemically defined anti-GAS vaccine [111]. Monoclonal and polyclonal antibodies directed against the cell-wall polysaccharide have been well characterized for binding to defined oligosaccharide fragments and to the corresponding glycoconjugates [25, 79,

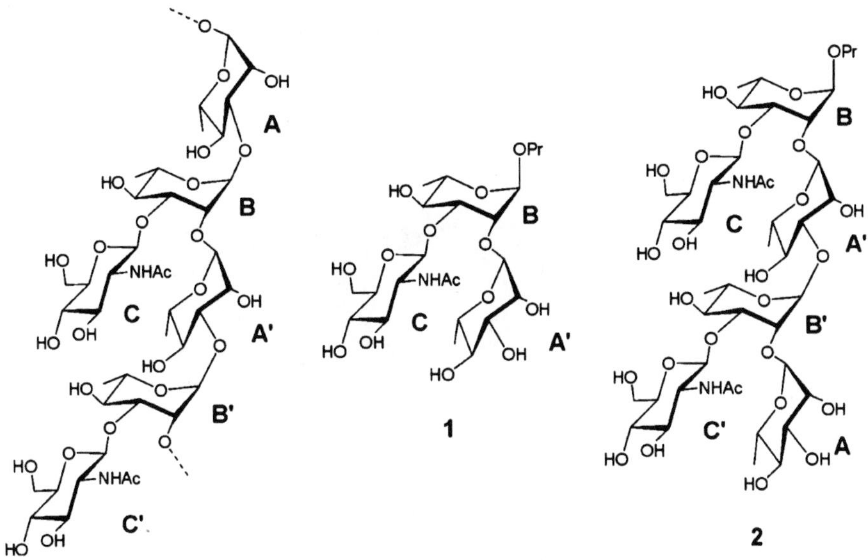

Scheme 4 Structures of the Group A *Streptococcus* cell-wall polysaccharide, and of branched trisaccharide (**1**) and hexasaccharide (**2**) fragments recognized by antibodies

112–117]. The interactions of carbohydrate-mimetic peptides with anti-GAS Abs have also been investigated.

The polysaccharide consists of a poly α-L-rhamnopyranosyl backbone with alternating (1-2) and (1-3) glycosidic linkages. Branching β-D-GlcNAc residues are attached at the 3-position of alternate rhamnose residues (Scheme 4) [118, 119]. GlcNAc is known to be the immunodominant monosaccharide component of the cell-wall polysaccharide, defined as the sugar that contributes the majority of binding energy [120]. However, it is important to note that the immunodominant sugar forms only part of a larger topographical surface [6, 120]. Immunochemical studies have indicated that the branched trisaccharide unit, Rha-α-(1-2)-(GlcNAc-β-(1-3)-)Rha (**1**, Scheme 4), is required for even moderate binding to most anti-GAS antibodies, and optimal binding is achieved by the extended topography presented by tetra- to hexasaccharide ligands [25, 79, 114, 117]. Correspondingly, although anti-GAS Abs may bind weakly to GlcNAc [25, 79, 121, 122], anti-GAS antibodies are not elicited by immunization with glycoconjugates comprising less than a pentasaccharide [117].

The branched trisaccharide **1** was defined as the minimal carbohydrate epitope for effective binding to anti-GAS Abs [79]. The bound conformation of **1** in complex with the IgG mAb Strep 9 was investigated by transferred NOE experiments and found to correspond to a local minimum conformation, which differed from the predominant conformation in the free state by a rotation around the ψ angle of the α-(1-2) linkage [123]. STD-NMR ex-

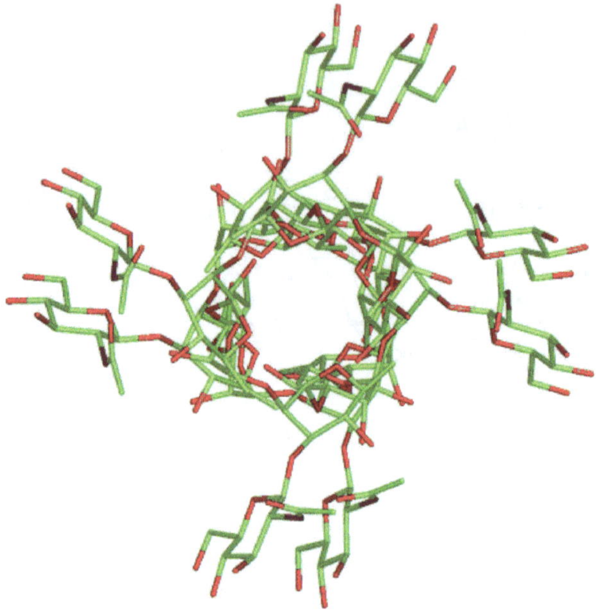

Fig. 7 A model of the cell-wall polysaccharide (CWPS) of group A *Streptococcus*, looking down the barrel of the helix. The core of the helix is composed of α-L-rhamnose residues with the β-D-GlcNAc moieties displayed on the periphery. Reproduced with permission from [99]. © 2004 Elsevier Science

periments to define the epitope revealed the strongest enhancements for the GlcNAc residue, with weaker enhancements for the Rha residues [124], consistent with a model of the interaction in which GlcNAc is accommodated in pockets within the combining site, while the Rha residues interact with the edges [79]. Similar patterns were obtained for a hexasaccharide, suggesting that the epitope comprises a surface stretching over two trisaccharide units, which interact in a similar way with the antibody. This would also be consistent with a helical model of the polysaccharide in which GlcNAc is presented on the outside of the helix (Fig. 7), and with its role as the immunodominant sugar forming part of a much larger epitope.

Carbohydrate-mimetic peptides for several anti-GAS antibodies were isolated by screening of a phage-display library [25]. The peptides isolated by the mAb Strep 9 have not been investigated, but a mimetic peptide for the mAb SA-3, an IgM mAb related to Strep 9, has been studied extensively. The peptide ADGADRPVPYGACG(Orn-biotin), bound to this IgM mAb with K_D 625 nM, while the hexasaccharide **2**, containing two repeating units, bound with K_D 17.5 μM (values determined by ITC). The binding of the peptide was more favorable entropically than that of the hexasaccharide. The binding of the hexapeptide DRPVPY, comprising the specific sequence isolated from the library without the flanking residues, was investigated by NMR spectroscopy.

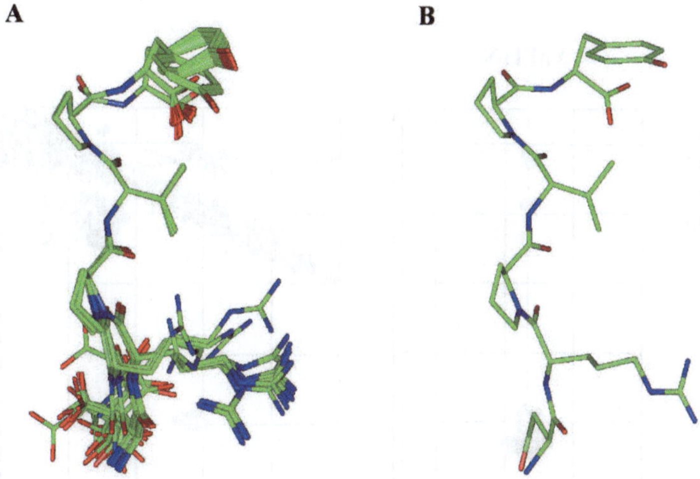

Fig. 8 Bound conformations of the hexapeptide DRPVPY. Superposition of structures was performed within Insight II and is based on the backbone atoms of residues 4–6. **A** Family of 27 structures calculated using trNOE-derived restraints only. **B** Average of the 27 structures in *panel A*. Reproduced with permission from [125]. © 2002 American Chemical Society

The K_D of the hexapeptide was estimated, from ^1H NMR line-broadening at different peptide/antibody ratios, to be 20 μM [125].

The conformation of the hexapeptide DRPVPY, when bound to this IgM mAb, was determined by transferred NOESY experiments [125]. The peptide adopts a tight turn conformation around the VPY residues, stabilized by hydrophobic contacts between the side chains (Fig. 8). Since the molecular weight of an IgM is roughly 1×10^6, the possibility of spin diffusion affecting the derived conformation was considered. The possibility for error caused by spin diffusion in transferred NOESY experiments has been investigated in both theoretical and experimental studies [126–135]. However, in many experimental studies, the extent of such error is not rigorously evaluated. Such evaluation is particularly important in studies with very large protein receptors, since the cross-relaxation rate in the bound state is increased [136]. To address this possibility, complete trNOESY buildup curves with mixing times from 50–200 ms were recorded, and the distance extrapolation method [137] was employed for distance estimation, in order to bias distances toward the shorter mixing times which would be less affected by spin diffusion. QUIET-trNOESY experiments allow the identification of cross-relaxation pathways mediated by spin diffusion [126, 127]. In this case, QUIET-trNOESY experiments revealed that one trNOE contact was possibly spurious (Fig. 9). Exclusion of this contact from the calculations did not significantly affect the resulting tight turn structure, since the network of other

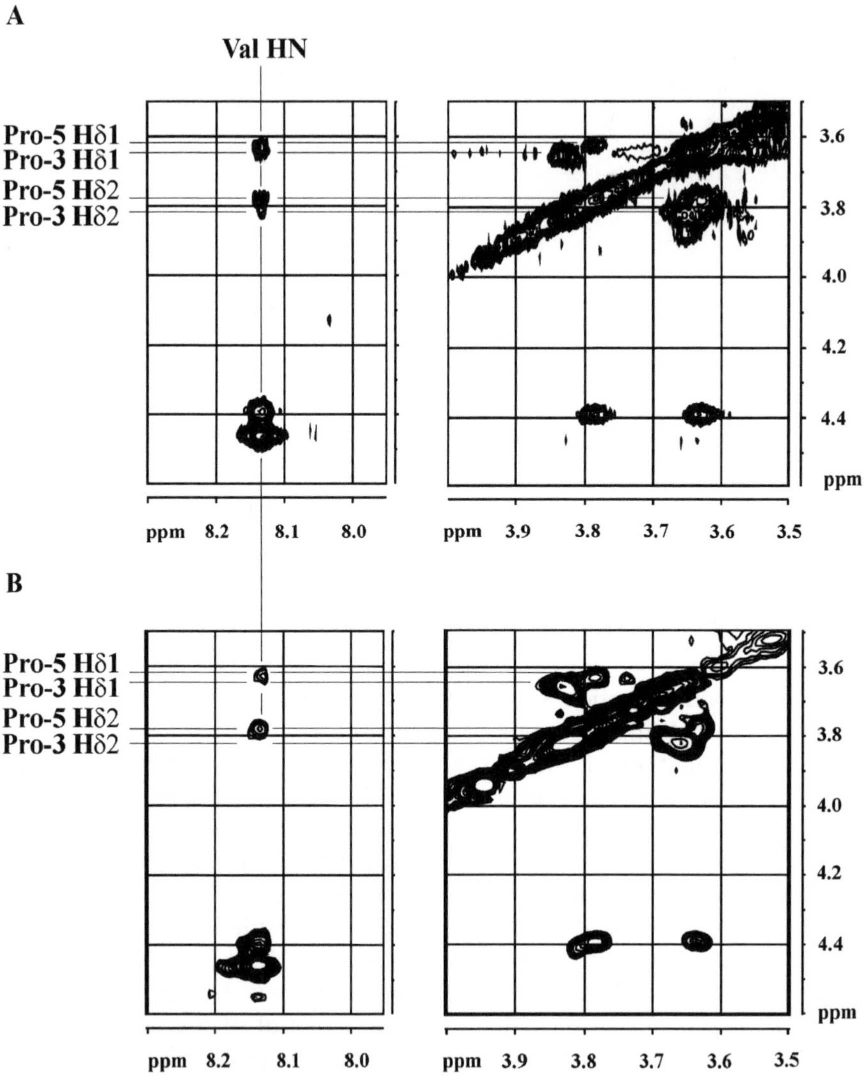

Fig. 9 TrNOESY and QUIET-trNOESY spectra of the peptide DRPVPY in the presence of the antibody SA-3. **A** Regions of a trNOESY spectrum ($\tau_m = 200$ ms) showing Val-4 HN–Pro-3,-5 Hδ cross-peaks and Pro-3,-5 Hδ–Hδ cross-peaks. **B** Corresponding regions of a QUIET-trNOESY spectrum ($\tau_m = 200$ ms) with inversion of 0.75 ppm wide bands centered at 4.15 ppm and 8.1 ppm; within the intersection of the quiet bands (quiet window), the Val HN–Pro-5 Hδ cross-peaks are still present, while the Val HN–Pro-3 Hδ_2 cross-peak is absent, indicating cancelation due to spin diffusion. Reproduced with permission from [125]. © 2002 American Chemical Society

trNOEs defining the turn was relatively dense. These results are encouraging for further investigations with IgM mAbs, which are often important in anti-carbohydrate immune responses.

STD-NMR experiments showed that all residues of DRPVPY except Asp 1 are in close contact with the antibody combining site—consistent with previous results in which RPXXY appeared as the consensus sequence from the phage-display library, and RP could not be replaced without loss of binding [25,102]. Furthermore, cleavage of the Tyr-6 residue also led to loss of binding, as assessed by the loss of trNOE contacts. Therefore, here the STD-NMR results correlate well with functional data, an observation which is also promising for the application of these NMR techniques to very large receptors. It should be noted also that the hexapeptide DRPVPY bound more weakly to the antibody than the longer peptide (14-mer) which contained flanking residues from the phage protein, indicating that some of these residues did contribute to binding. This observation is perhaps not surprising considering that the binding site is likely similar to the related mAb Strep 9, for which the carbohydrate epitope comprises an extended surface, and molecular models suggest a long, groove-shaped binding site [79].

There is some evidence for the population of a turn conformation similar to the bound conformation in the free peptide, however this is expected to be only one of many conformations adopted by the free peptide, as other NMR parameters (chemical shifts and $^3J(H^N–H^\alpha)$ values) indicated conformational averaging, as is observed for many short peptides. The presence of a certain amount of the bioactive conformation suggested that this peptide could potentially be a good mimotope. Immunization experiments in mice, using the DRPVPY hexapeptide conjugated to tetanus toxoid, demonstrated a high titer anti-carbohydrate immune response, showing that the peptide was indeed an effective mimotope [138].

5
Mimicry of a Complex Fungal Capsular Polysaccharide

Cryptococcus neoformans is an opportunistic fungus that often infects immunocompromised patients. The organism is encapsulated with a glucuronoxylomannan polysaccharide, which is an important antigen of the fungus, but has limitations in vaccine development [139]. As is typical of polysaccharides, it is a T-cell independent antigen that does not induce affinity maturation, class switching and memory response; and in addition, polysaccharide–protein conjugates may elicit not only protective antibodies but also nonprotective and disease-enhancing antibodies, leading to an ineffective immune response [140]. The chemical nature of the oligosaccharide epitopes is currently not completely defined; additional modifications such as variable O-acetylation add further complexity [141]. Carbohydrate-mimetic peptides are considered a promising strategy to circumvent these complications, with the goal of obtaining a solely protective, anti-polysaccharide immune response.

MAbs directed against the *C. neoformans* polysaccharide are closely related, and some differ only in a few amino acid residues [142, 143]. This fine specificity is related to their protective efficacy. A highly protective IgG1 mAb was used to screen phage-displayed peptide libraries, and several binding peptide sequences were identified [144]. The dodecapeptide GLQYTPSWMLVG bound to the mAb with a K_D of 295 nM, measured by surface plasmon resonance (SPR); and inhibited binding of the mAb to the polysaccharide. The dodecapeptide was able to induce only a weak anti-polysaccharide antibody response [145].

Some possible reasons for the inability of this dodecapeptide to act as an effective mimotope were revealed by the crystal structure of this peptide in complex with the protective mAb (Fig. 10) [146]. The dodecapeptide forms two tight turns in the antibody combining site. A type II β-turn (Ser-7 to Leu-10) and inverse γ-turn (Leu-2 to Tyr-4) allow the placement of almost all the amino acid side chains into the binding site. The majority of contacts are made by the residues Tyr-4 to Leu-10, which comprise the consensus sequence isolated from the phage-display library. The Met-9 side chain is buried deeply within the site. There are signs of a nonoptimal fit between the peptide and the antibody: two hydrophobic cavities are present, where the peptide does not contact the binding site. The peptide makes few contacts to the antibody heavy chain, which is often important in conferring antigen specificity [147], and would likely be important in inducing an effective immune response. The majority of intermolecular interactions are van der Waals contacts, in keeping with the observation that the combining site is predominantly hydrophobic. Although the structure of the complex between the antibody and the glucuronoxylomannan is unknown, it is suspected that structural mimicry is not the case for this particular peptide.

A novel strategy was applied to investigate the similarities and differences in antibody recognition of the polysaccharide and peptide ligands. Two monoclonal anti-*Cryptococcus* mAbs, one protective and one nonprotective, were shown to differ by only two amino acid substitutions (these mAbs were closely related, although not identical to the structurally characterized mAb described above) [140]. The protective mAb was subjected to site-directed mutagenesis to replace these two amino acid residues with those of the nonprotective mAb. As expected, this resulted in a loss of protective efficacy, and a loss of binding affinity for the polysaccharide. However, these replacements did not confer binding to a 14-residue peptide that was normally recognized only by the nonprotective mAb. Thus, polysaccharide and peptide recognition within the binding site of this mAb showed some differences. In addition to this double replacement, several other single mutations of residues within the binding site caused changes in the pattern of peptide recognition that differed

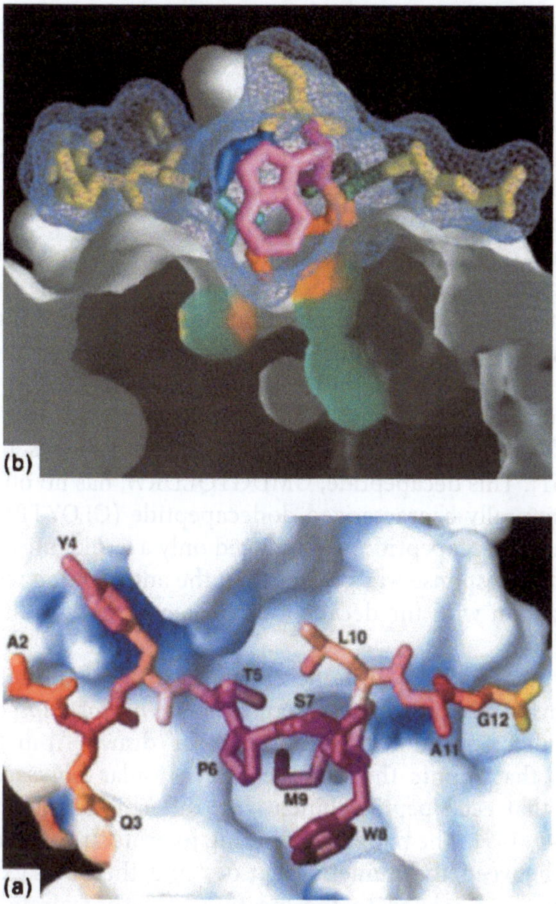

Fig. 10 **a** Electrostatic surface of the antibody 2H1, where positively charged regions are *blue*, and negatively charged regions are *red*. The peptide is colored according to its *B*-factors, with *dark purple* for the lowest *B*-factors (41 Å^2 to 42 Å^2), going to *yellow* for the highest (54 Å^2 to 55 Å^2). Peptide residues are labeled using their one-letter amino acid code. **b** Cutaway molecular surface of 2H1 shown in *white*, side view. The peptide is in *yellow*, except for the motif residues, which are colored as follows: Thr^{P5} *dark blue*; Pro^{P6} *light blue*; Trp^{P8} *pink*; Met^{P9} *orange*; Leu^{P10} *green*. The molecular surface of the peptide is shown as a *lilac* mesh. Also shown are the surfaces of the two cavities that remain between the bound peptide and the antibody-binding surface. These surfaces are colored *yellow*, *green*, and *orange* according to their proximity to the antibody light chain, heavy chain, or peptide, respectively. The orientation of the peptide is similar to that in (**a**) with Ala^{P2} on the *left*, Gly^{P12} on the *right*. Reproduced with permission from [22]. © 1997 Elsevier Science

from the changes observed for carbohydrate recognition. Several antibody residues important in binding both peptide and carbohydrate were identified, while the pattern of interactions with other residues in the binding site was clearly different. This study illustrates a novel approach which begins to define requirements for Ab–ligand molecular recognition even in the absence of any structural information or knowledge of the nature of the oligosaccharide epitope, and provides additional evidence that peptide and carbohydrate recognition by mAbs is not based on exact structural mimicry. The study also provides another example of the use of peptides to probe differences between closely related mAbs, which cannot be accessed, for one reason or another, with oligosaccharide probes (see also [20, 25]).

More recently, the *Cryptococcus* mimetic peptides were used to show differences in ligand recognition which depended solely on isotype between V-region identical antibodies, a groundbreaking result as Ab/ligand molecular recognition was generally considered to depend only on the V-region of the Ab [148].

Another study using a human anti-*Cryptococcus* mAb was successful in isolating an effective peptide mimotope of the *C. neoformans* glucuronoxylomannan [60, 61]. This decapeptide, GMDGTQLDRW, has no obvious similarity to the structurally characterized dodecapeptide (GLQYTPSWMLVG) described above. The decapeptide also induced only a weak anti-polysaccharide response, but this response was greater than the anti-peptide response generated. Immunization with the decapeptide protected mice against *Cryptococcus* infection [60].

Clearly, further studies will be necessary to sort out the multiple factors involved in the in vivo immune response to *C. neoformans* carbohydrate-mimetic peptides. Several conclusions may be drawn from the results to date. Peptides that mimic the cryptococcal capsular polysaccharide show specificity, in that each peptide binds with differing affinity to closely related mAbs [140, 149]. The pattern of binding to protective and nonprotective mAbs differs between the mimetic peptides and the polysaccharide [140]. Protective efficacy is related to the location of carbohydrate epitopes recognized by these mAbs, within the polysaccharide capsule, but likely also depends on interactions between mAbs and cellular responses [149]. Peptides have been shown to be functional, immunogenic mimics, but their protective efficacy depends on multiple factors, including the type of Abs elicited and interactions with the cellular immune system. Protective efficacy does not correlate with binding affinity to representative mAbs, but rather depends on the nature of these interactions.

6 Peptide Mimics for Mannose-Recognizing Proteins – Interactions with Different Receptors

One of the earliest reported carbohydrate-mimetic peptides was the dodecapeptide DVFYPYPYASGS [10]. The peptide acts as a ligand for concanavalin A, a lectin which binds oligosaccharides bearing terminal α-linked mannose or glucose residues. It is tetrameric, with a carbohydrate binding site on each monomer, and binds multivalently to oligosaccharides displayed on cell surfaces. The dodecapeptide was found to bind with similar affinity to that of methyl α-D-mannopyranoside, and could compete with methyl α-D-mannopyranoside for binding. Functional studies showed a complicated picture in which the dodecapeptide and carbohydrate were also shown to cross-react immunologically [151, 152]. Thus, polyclonal antibodies raised against a peptide–protein conjugate bound to methyl α-D-mannopyranoside, and vice versa; and the peptide inhibited concanavalin A-induced T cell growth [153].

The YPY motif of this peptide was identified as a consensus sequence in phage library screening [10], and is also important for concanavalin A binding. The crystal structure of the dodecapeptide in complex with concanavalin A [153] revealed that the observed molecular mimicry was not based on structural similarity between YPY and saccharides. The peptide did not bind at the sugar binding site, but at a different site nearby. In fact, two separate peptide-binding sites were observed, with the peptide bound in two different conformations involving different interactions with the protein (Fig. 11). The first binding site is a shallow crevice on the protein surface of one monomer, while in the second binding mode, the peptide was found between two crystallographically related monomers. Some conformational differences between the protein monomers were observed, and the second binding site appeared to be influenced by crystal contacts. The peptide did

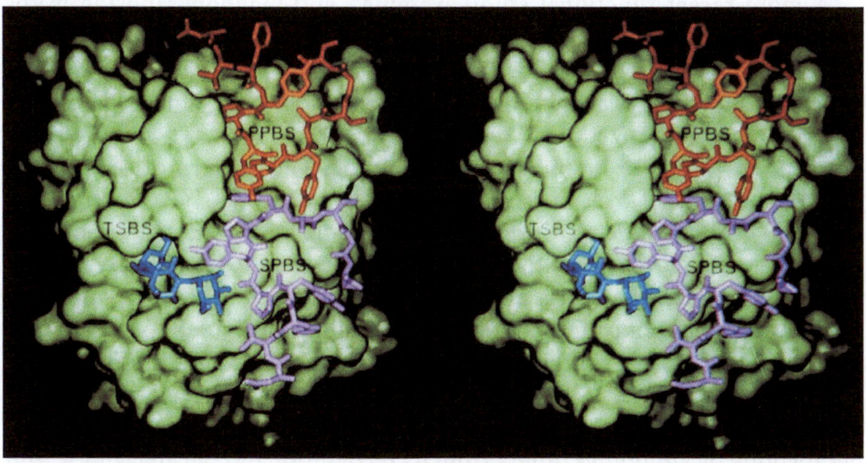

Fig. 11 A stereoview of the A subunit of concanavalin A. The dodecapeptide DVFYPYPYASGS is shown bound in the primary peptide-binding site (PPBS; colored *orange*), and in the secondary peptide-binding site (SPBS; colored *pink*). The secondary peptide-binding site also involves extensive contacts with another crystallographically related conA subunit (not shown). For comparison, the location of the trisaccharide-binding site (TSBS), with bound trimannose (*blue*), is also shown. Reproduced with permission from [153]. © 2000 The American Society for Biochemistry and Molecular Biology

not extend into the trimannoside binding site, although there was a small amount of overlap with the pentasaccharide binding site [154]. These results show that mimicry in this system is functional and not structural. However, the carbohydrate binding site was bound up in crystal contacts in the peptide complex and inaccessible to the peptide, and the possibility remains that the peptide might bind at this site in solution.

The crystal structures of two related peptides of different lengths in complex with concanavalin A showed similar interactions as the first peptide, with some conformational flexibility depending on length, and greatest similarity of the conserved YPY motif [155]. A hexapeptide, acetyl-MYWYPY-amide, with greater binding affinity, was designed based on these crystal structures [156], and bound within the primary peptide binding site with improved shape complementarity. However, the peptide conformation varied depending on which concanavalin A monomer bound the peptide, and significant differences were observed with respect to the longer peptides, even within the conserved YPY sequence. Overall, the peptides bound in this site showed a great deal of conformational variation. A molecular modeling study suggested that the dodecapeptide could possibly bind within the carbohydrate-binding site, and vice versa; these results raised the possibility that the peptide-binding sites actually formed part of an extended carbohydrate-binding site, so that the observed biochemical effects might result from partial overlap of the binding sites [157]. This hypothesis remains to be investigated further.

A more recent functional study examined anti-mannopyranoside mAbs generated in mice. Of 12 individual anti-mannoside mAbs, seven showed significant binding to the dodecapeptide described above [158]. Using small-scale peptide synthesis on pins, peptides with each residue replaced by glycine were synthesized, and the residues Phe-3, Tyr-6, Tyr-8, Ser-10 and Ser-12 were shown to be important in binding. This pattern of recognition differs from that of concanavalin A, in which mainly the YPY motif is involved in binding. SPR measurements were performed to measure affinities, which were similar between the peptide and α-D-mannopyranoside. Further competition assays showed a new observation, that the peptide-recognizing mAbs were not specific for mannose—they bound equally well to glucose and lactose, and any one of these three sugars was able to inhibit binding to the dodecapeptide.

Taken together, these results suggest that molecular recognition of the dodecapeptide by antibodies differs from its recognition by concanavalin A, and that the immunological cross-reactivity observed in some studies does not reflect structural mimicry. That molecular recognition by concanavalin A of carbohydrates and peptides differs has also been shown in structural studies. Here, the functional molecular mimicry observed with respect to different receptors should not be assumed to imply structural mimicry—the intermolecular interactions may differ in each case.

7 Molecular Modeling Studies

The origin of peptide–carbohydrate mimicry has been addressed in several systems by molecular modeling. Since the computational prediction of the binding modes of biomolecules is challenging, and with current methods may not always be accurate, such models should be considered with some degree of uncertainty, and confirmation by independent experiments (mutagenesis of the receptor or ligand; or synthesis of analogs) is usually desirable. However, when experimental structure determination is not possible, theoretical studies can provide much insight into molecular recognition, and the reliability of docking calculations can be increased by the inclusion of suitable controls (e.g. ligands with known binding affinity). Some recent examples are described below.

7.1 Concanavalin A

The crystal structure of concanavalin A in complex with a pentasaccharide [154] was used as a starting point to identify new peptide ligands. The peptide sequence RYGRY was identified by molecular modeling using the program LUDI [159], in which fragments of a ligand are placed in favorable positions within the receptor binding site, and then used to build the complete ligand. In the modeled RYGRY complex, the peptide contacted two of the same residues as the pentasaccharide, Leu-99 and Tyr-12, while the rest of the interactions were different. Functional mimicry was observed, in that binding of concanavalin A to multivalent (MAP) forms of this peptide was inhibited by methyl α-D-mannopyranoside. The K_D values measured by ITC were similar to those of carbohydrate ligands [160]. Immunization of mice with the MAP peptide induced polyclonal antibodies that neutralized HIV in cell culture, presumably due to binding to mannose-containing oligosaccharides expressed on the gp120 protein. The computational model predicts that this functional mimicry is not based on structural similarity: the peptide is predicted to bind within the carbohydrate-binding site, but not to act as a structural mimic of the carbohydrates, as most of the ligand–receptor contacts in the predicted peptide complex differ from those of the carbohydrate.

7.2 Anti-Lewis Y Antibodies—Molecular Modeling and Functional Data

Molecular modeling has been applied to investigate peptide mimicry of the Lewis Y tetrasaccharide. The peptide APWLYGPA, first reported by Hoess et al. [12], and shown to be a mimotope [54], binds to the monoclonal anti-Lewis Y antibody B3. A model of the antibody was constructed computationally, by mutation from the crystallographic structure of the homologous antibody BR96, followed by energy minimization [161]. The complex of the modeled antibody with the Le^Y tetrasaccharide was constructed, based on knowledge of conserved hydrogen-bonding residues within the site. The LUDI program was applied to determine the best placement of fragments corresponding to the previously identified peptide sequence APWLY. In the resulting model, the peptide forms a β-turn comprising residues Pro-2 to Tyr-5. A comparison of the modeled peptide and carbohydrate complexes revealed one similar interaction: the Ala-1 side chain in the peptide complex occupies a similar position as the GlcNAc *N*-acetyl methyl group of Le^Y. The remaining interactions differ between peptide and carbohydrate.

Another monoclonal anti-Le^Y antibody, BR55, does not bind to APWLY-GPA, but does recognize peptides containing the motifs YRY and WRY [49]. When presented in multivalent form, peptides containing these motifs inhibited the binding of mAb BR55 to solid-phase, multivalent Le^Y; both had IC_{50} values of 1.6 μM. Comparative modeling of this antibody was used to generate a model of its complex with Le^Y [162] and with peptides [55]. A LUDI search was used to identify a novel peptide ligand, FSLLW. The predicted binding mode included a few structurally mimetic interactions supplemented by many other favorable interactions. This peptide sequence was subsequently selected by phage library screening experiments, providing support for the computational model. Immunization with multivalent forms of the peptide led to an anti-Le^Y immune response, which was more specific than that induced by immunization with Le^Y, lending support to the hypothesis that peptides may discriminate between subsets of anti-carbohydrate antibodies more effectively than the natural carbohydrate antigens [25, 55].

7.3 The Receptor for Hyaluronan-Mediated Motility (RHAMM)— Mimicry by Charged Groups

The receptor for hyaluronan-mediated motility (RHAMM) mediates cell migration and proliferation, and is implicated in metastasis and overexpressed in tumor cells. Binding to extracellular matrix hyaluronan (poly-(β-(1-4)-GlcA-β-(1-3)-GlcNAc)-) is required to initiate the cellular signaling pathways leading to activation of these functions. Peptide ligands for RHAMM were identified from one-bead, one-peptide libraries, with one library containing random octapeptide sequences, and another which was biased to present acidic residues (Asp or Glu) at every second position; this arrangement was designed to mimic the spacing of the carboxylate groups of the glucuronic acid residues in hyaluronan [36]. Both L- and D-configurations of Asp and Glu were included, as the presence of D-amino acids was predicted to provide a closer spatial resemblance to the global minimum conformation of hyaluronan, if an extended conformation for the peptide was assumed. Several peptide sequences with strong binding affinity were identified from both libraries (for example, CYDSeYeSe, K_D 8 nM; CEGEWPVYP, K_D 48 nM). These peptides could find applications in cancer treatment and in drug targeting to tumors, and would have the advantage of resistance to degradation by hyaluronidase.

Computational docking was employed to predict the binding modes of these octapeptides within the hyaluronan binding site, using an NMR-based model of the HA-binding domain of RHAMM [163]. Solvent-exposed, positively charged lysine side chains are spaced 11 Å apart in the binding site and thought to interact with the carboxylate groups on hyaluronan. These interactions, along with hydrophobic interactions with two exposed valine residues, are predicted to be responsible for hyaluronan binding. Docking predicted that peptide binding would involve electrostatic interactions of the acidic residues of the peptides with these Lys residues, but no corresponding hydrophobic interactions, and the peptides were also predicted to make additional contacts to other basic residues. Multiple binding modes were predicted for the multiple peptide sequences studied. This study provides another example of peptide binding involving a mixture of structural mimicry and additional, favorable interactions with the site. The predicted extensive involvement of charged residues in structural mimicry is a novel feature of this system.

7.4 Ganglioside GD2

Gangliosides are expressed on the membranes of normal mammalian cells, where they function in differentiation, but their expression patterns are altered in cancer cells. The disialoganglioside GD2 (GalNAc-β-(1-4)-(NeuAc-α-2-8)-NeuAc-α-(2-3))-Gal-β-(1-4)-Glc-β-(1-1)-Cer) is a tumor-associated antigen which is expressed on neuroblastoma cells. A mouse-human chimeric antibody, ch14.18, directed against GD2, is used in passive immunotherapy to reduce residual disease. This mAb was used to identify carbohydrate-mimetic peptides [44]. The X-ray crystal structure of the Fab fragment of ch14.18 is known [164] and molecular modeling was used to predict the binding mode of GD2 within the combining site. The binding mode is predicted to depend mainly on hydrogen bonding, and shows one NeuAc residue and the GalNAc residue forming hydrogen bonds to polar residues such as Thr, Asn, Ser and Tyr. The branching Gal residue lies outside the binding site, as does the α(2-8)-linked NeuAc residue, while most contacts are made by the terminal residues [164].

Molecular modeling was also used to predict the binding modes of two mimetic dodecapeptides, CDGGWLSKGSWC and CGRLKMVPDLEC, using the coordinates of the Fab X-ray structure [44]. Docking was performed with AutoDock version 3.0 [165], employing a Lamarckian genetic algorithm and allowing side chains of the peptides to be flexible. The predicted binding

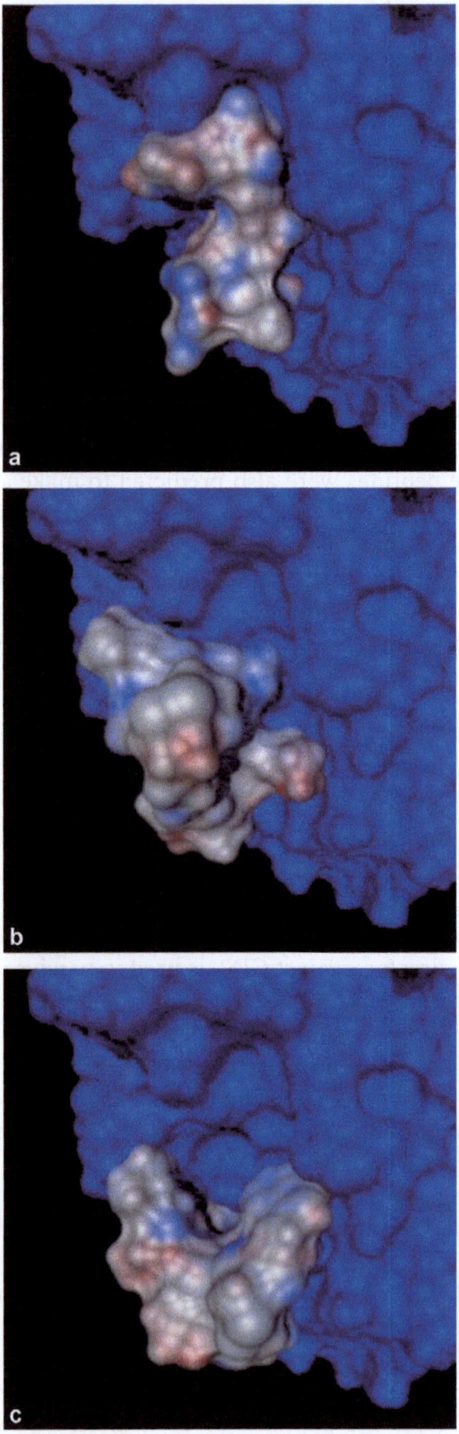
a
b
c

◀ **Fig. 12** Predicted binding modes of the GD2 ganglioside and two peptide mimics, in complex with the anti-GD2 Fab fragment. **A** Binding of the GD2 ganglioside to the Fab fragment. The structure of the Fab was taken from the theoretical model 2PSK. **B** Binding of the CGRLKMVPDLEC peptide to the Fab fragment. **C** Binding of the CDGGWL-SKGSWC peptide to the Fab fragment. The antibody and ligands are shown in molecular surface representation, with the antibody surface *blue* and the ligand surfaces colored according to electrostatic potential. Reproduced with permission from [44]. © 2005 Elsevier Science

modes differed for each peptide. Neither was able to fill the entire groove as had been predicted for the ganglioside; rather, each peptide filled roughly half the pocket and interacted additionally with the edges of the binding site (Fig. 12). Some key residues (His H35, Arg L90) thought to be important in ganglioside binding were also contacted by the peptides [44].

8
Peptide–Carbohydrate Mimicry in Enzyme Active Sites

To date, the majority of carbohydrate-mimetic peptides that have been isolated interact with antibody receptors. However, examples of peptide ligands for carbohydrate-processing enzymes also exist. Eichler et al. [19] described the selection of cyclic peptides which acted as high-affinity α-glucosidase inhibitors from phage display libraries. Taki et al. [23] isolated peptide ligands for an antibody directed against lactotetraosylceramide, which were also found to modulate β-galactosidase activity; enzyme activation and inhibition were both observed, under different conditions. Interesting examples of naturally occurring peptide–carbohydrate mimicry are provided by the crystal structures of two cyclic peptide chitinase inhibitors in complex with family 18 chitinases [166]. Here, as in the case of the anti-*Shigella* antibody, crystal structures of the receptor in complex with both peptide and carbohydrate ligands are available, allowing a direct examination of the nature of mimicry.

8.1
GlcNAc$_5$-ChiB Complex

The crystal structure of a pentamer of GlcNAc residues, representing the chitin polymer (poly-β-(1-4)-GlcNAc), bound to the chitinase enzyme ChiB from *Serratia marcescens*, revealed a narrow, tunnel-like active site in the center of the TIM barrel fold [167]. Several conserved residues near the center of the site, which are important in catalysis, interact with the substrate via hydrogen bonds, while interactions farther from the center depend on van der Waals interactions. The sugar in the – 1 subsite adopts a ^{1,4}B boat conformation, presumably due to interactions with these critical active-site residues.

In this conformation, the *N*-acetyl carbonyl oxygen is found only 3.0 Å from the anomeric carbon, in a position favorable for nucleophilic attack on C1 and formation of an oxazolinum ion, which is the first step in the substrate-assisted catalytic mechanism of family 18 enzymes.

Some key interactions included a hydrogen bond from the GlcNAc carbonyl oxygen to the side chain hydroxyl group of Tyr 214; and hydrogen bonds from the GlcNAc amide proton to the side chain of Asp 142. Binding of the $GlcNAc_5$ substrate was associated with closing of the "roof" of the active site tunnel; with movement of Trp 97 and Trp 220 toward the center of the site; and with movement of another important catalytic residue, Asp 142, which hydrogen bonds to Asp 140 in the unliganded enzyme, but in the complex hydrogen bonds to Glu 144. It was proposed that this promotes a proton transfer from Glu 144 to the glycosidic oxygen.

8.2 Argifin–ChiB Complex

The structures of the complexes of argifin and argadin with ChiB allowed a comparison with the $GlcNAc_5$ complex and description of the nature of mimicry in this system [166]. Figure 13 shows the chemical structures of these cyclic peptides, which were isolated from fungal cultures [168, 169].

The argifin ligand occupies a space within the active site roughly corresponding to $GlcNAc_3$ bound in subsites – 1, + 1 and + 2. The cyclic portion of argifin is found at the nonreducing subsites, while the arginine "tail" extends into the – 1 subsite (Figs. 14, 15). The tail mediates contacts with the critical catalytic residues, including some contacts which mimic those of the carbohydrate substrate. The two guanidyl amide nitrogens act as hydrogen bond donors to the carboxylate oxygens of the catalytic base Glu 144; the *N*-methyl-carbamoyl amide nitrogen donates a hydrogen bond to the side chain of Asp 142; and the *N*-methyl-carbamoyl carbonyl oxygen accepts a hydrogen bond from the side chain hydroxyl group of Tyr 214. These interactions exactly mimic those of the glycosidic oxygen and the *N*-acetyl group of the GlcNAc residue in the – 1 subsite in the substrate complex, with the exception that the *N*-acetyl amide nitrogen donates a bifurcated hydrogen bond to Glu 144 and Asp 142, whereas the peptide provides two separate hydrogen bonds.

The *N*-methyl-carbamoyl methyl group is buried in a hydrophobic pocket of residues Tyr 10, Ala 184, and Met 212, which is designed to accommodate the *N*-acetyl methyl group of GlcNAc. The atoms described above are also found in very similar positions within the active site in the two complexes: the C and NH of the guanidyl group occupy identical positions to C1 and O5 of GlcNAc, the other NH of the guanidyl group occupies an identical position to the *N*-acetyl NH of GlcNAc, and the other NH occupies the position of O4. A water-mediated hydrogen bond from the guanidyl NH1 to Arg 294 replaces that of O5 of GlcNAc, with the water being found near the position of O5.

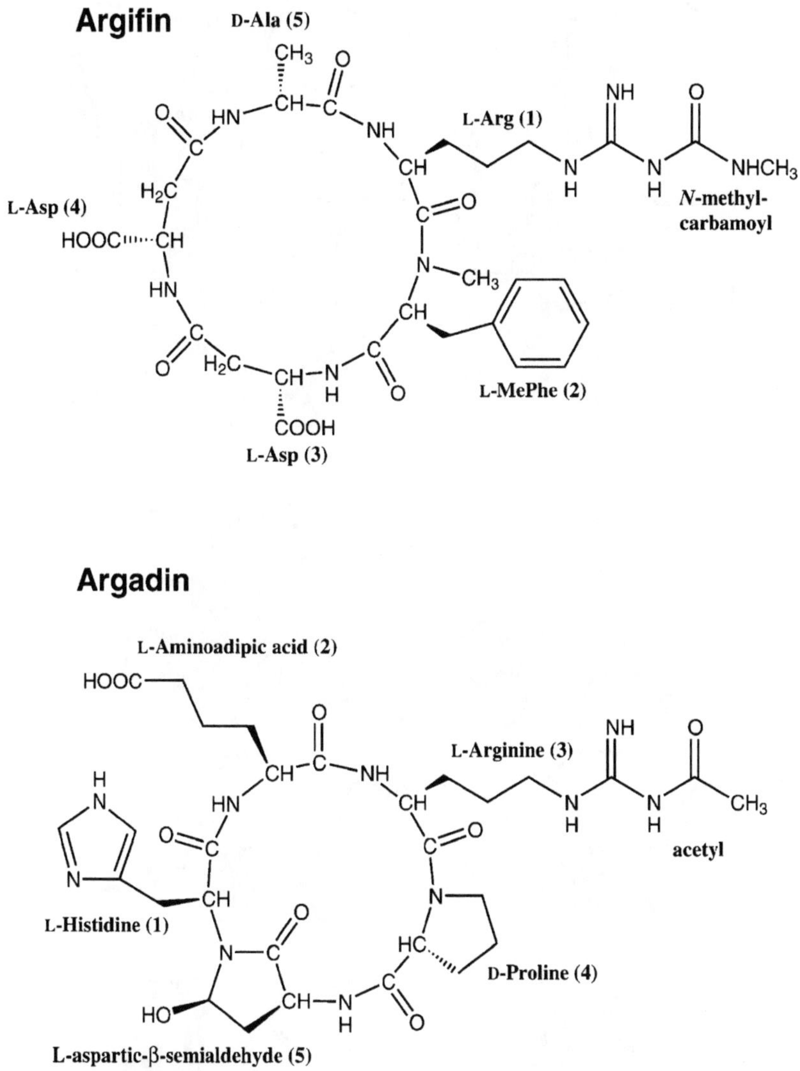

Fig. 13 The chemical structures of argifin and argadin. Reproduced with permission from [170]. © 2005 Elsevier Science

Finally, the CG and CD atoms of the arginine residue of the tail occupy the positions of C3 and C4 of the + 1 GlcNAc residue. Thus, this arrangement may be considered structural mimicry, and one side of the – 1 sugar residue as well as part of the + 1 sugar residue are mimicked effectively by the peptide.

The remaining portion of the – 1 residue (C3-C6 and O3-O6) is not mimicked by the peptide. This face of the ring contacts Trp 403 through van der Waals interactions and hydrogen bonds from O7 to Trp 403 NE1 and Gln 407 NE2, and the peptide does not make corresponding contacts. The – 2 sugar

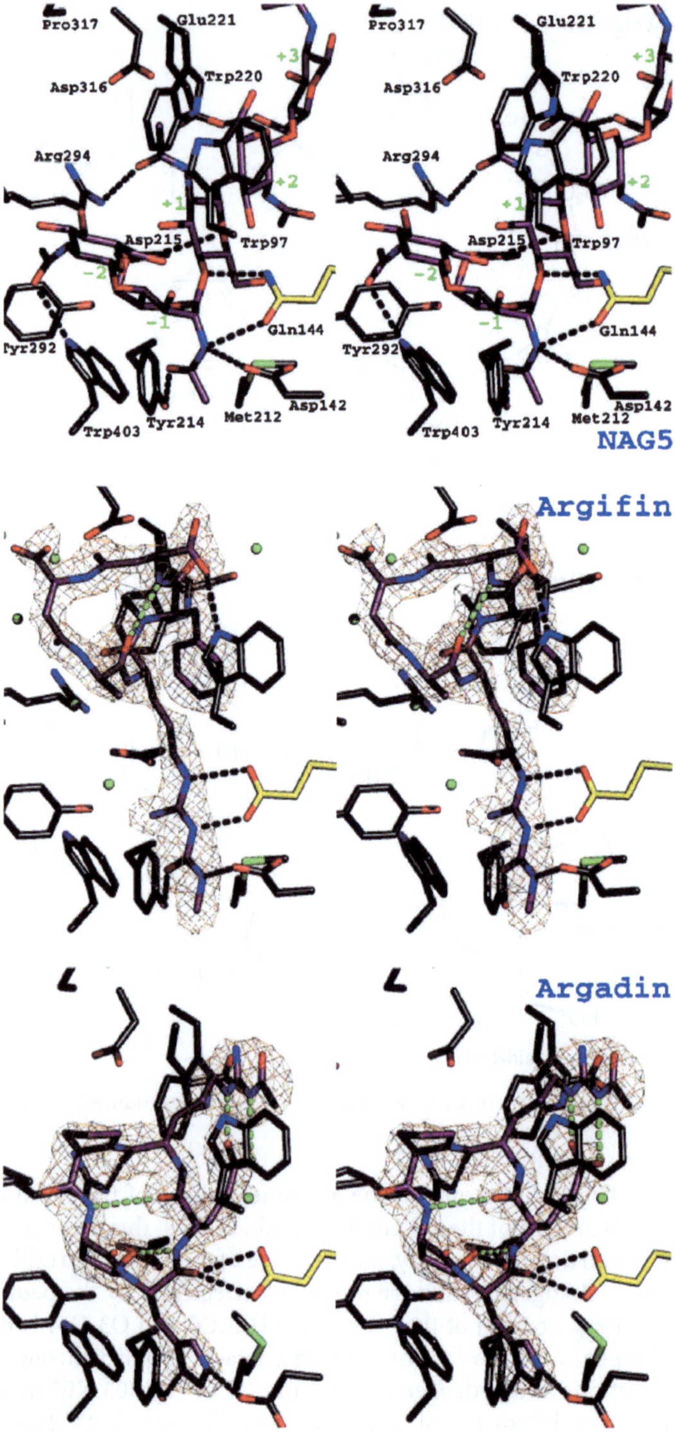
Pro317
Glu221
Asp316
Trp220
+3
Arg294
+2
+1
Asp215
Trp97
-2
-1
Gln144
Tyr292
Asp142
Met212
Trp403
Tyr214
NAG5
Argifin
Argadin

◀ **Fig. 14** Stereoviews of the active site of *Serratia marcescens* ChiB with bound $GlcNAc_5$ ([167]; the receptor is mutationally inactivated ChiB, where the catalytic Glu-144 has been replaced with a glutamine) and with bound argifin and argadin ([166], where the receptor is the wild-type ChiB enzyme). Unbiased (i.e., before including any inhibitor model) $|F_o - F_c|$, Φ_{calc} (contoured at 2.5σ) maps are shown in *orange*. Ligand carbon atoms are colored *purple*. Side chains interacting with the cyclopentapeptides are shown in a stick representation, with carbons colored *gray*, except for the catalytic residue 144, for which carbons are shown in *yellow*. Tyr-145 (which only hydrogen-bonds to $GlcNAc_5$) has been omitted to improve clarity. Water molecules hydrogen-bonding to both protein and inhibitor are shown as *green spheres* (hydrogen bonds are not shown). Hydrogen bonds between the ligands and the protein side chains are shown as *black dotted lines*. Argifin/argadin intramolecular hydrogen bonds are shown as *green dotted lines*. In the complex with $GlcNAc_5$, the sugar subsites are indicated by *green labels*. Reproduced with permission from [166]. © 2002 by the National Academy of Sciences of the USA.

occupies a relatively hydrophobic subsite which includes Trp 403, Tyr 292, and Phe 12; the peptide does not occupy this area, and ordered water molecules fill this subsite in the peptide complex.

In the + 1 subsite, GlcNAc is oriented with C-6 near the bottom of the site in contact with Met 212, and with the *N*-acetyl group at the top, interacting with Trp 97. O6 forms a hydrogen bond to the hydroxyl group of Tyr 145. In the peptide, the backbone and $C\alpha/C\beta$ atoms of Arg 1 are found in this area and also make contact with Trp 97. The peptide does not extend completely to the bottom of the site, and does not have corresponding interactions with Met 212 or Tyr 145. The + 2 sugar ring makes van der Waals contacts with Trp 220, while its *N*-acetyl group is found in a pocket of Gly 188, Leu 265, and the backbone of Leu 216. Again, the peptide does not extend deeply into the site in this area. Rather, the side chain of the *N*-methylphenylalanine residue bridges the + 1 and + 2 subsites by making contact with both Trp 97 and Trp 220. This is associated with a slight widening of the site, with an increased distance between the two Trp residues on opposite sides. The + 3 subsite is not contacted by the peptide at all.

Conversely, the peptide has interactions with the top and roof of the active site, and in fact binding causes some opening of the roof: the side chain of Ala 3 and the backbone of N-MePhe 2 both contact Trp 97, and residues 2, 4 and 5 contact Asp 316 and Pro 317, which form part of the "roof". The side chain of Asp 316 is directed away from the active site in order to accommodate the bound peptide, and the backbone of the loop containing residues 316–317 moves also.

Overall, the peptide provides a subset of structurally mimetic interactions, especially through the *N*-methyl-carbamoyl tail in the – 1 subsite. However, interactions with the rest of the binding site differ, and peptide binding leads to widening of the binding site and opening of the roof, as opposed to carbohydrate binding, which leads to narrowing of the binding site and closing of the roof.

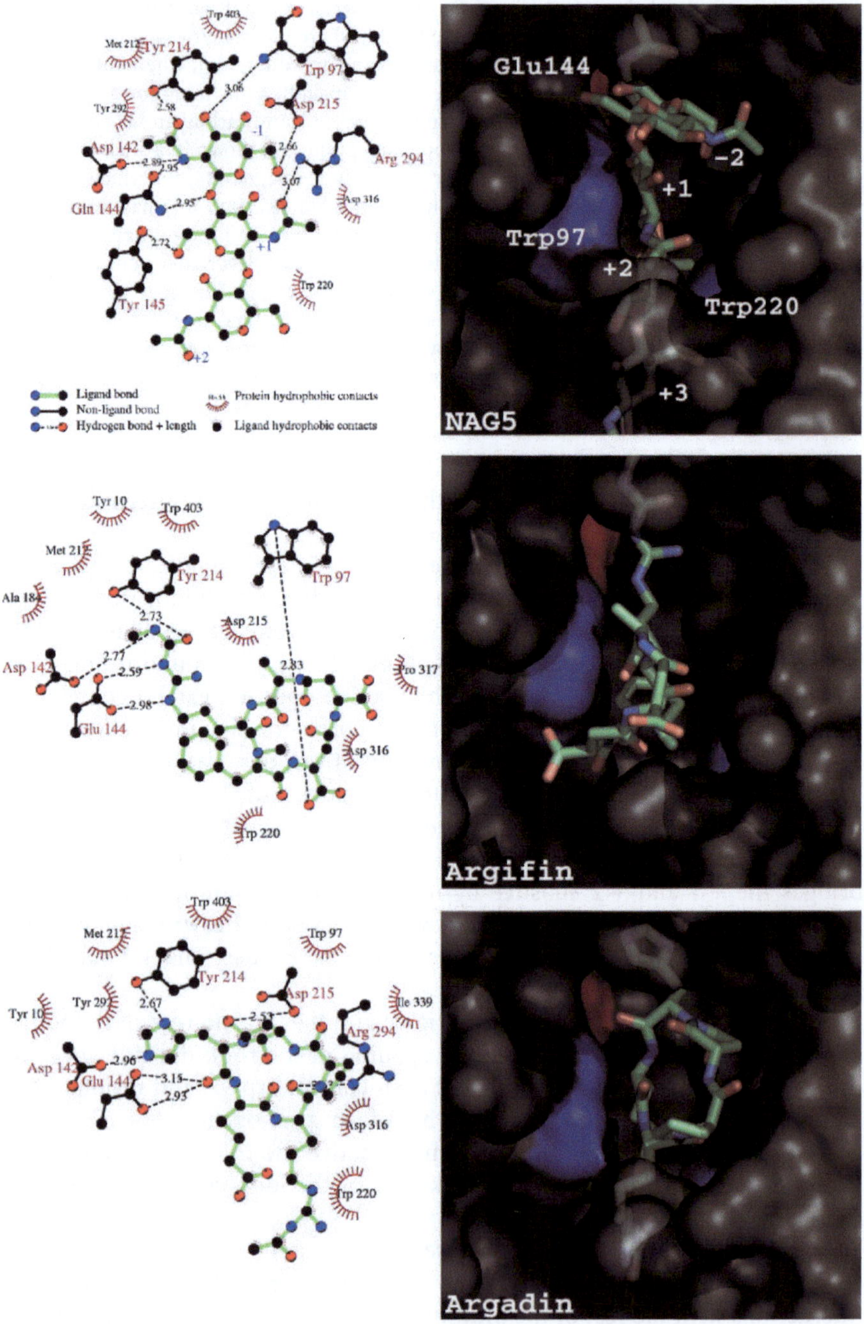
Trp 403
Met 212
Tyr 214
Trp 97
Tyr 292
sp 215
Asp 142
-1
Arg 294
Gln 144
Asp 316
+1
Trp 220
Tyr 145
+2
Ligand bond
Non-ligand bond
Hydrogen bond + length
Protein hydrophobic contacts
Ligand hydrophobic contacts
Glu144
-2
+1
Trp97
+2
Trp220
+3
NAG5
Tyr 10
Trp 403
Met 21
Tyr 214
Trp 97
Ala 18
sp 215
Asp 142
Glu 144
ro 317
sp 316
rp 220
Argifin
Trp 403
Met 21
Trp 97
Tyr 214
sp 215
Tyr 10
Tyr 29
le 339
Arg 294
Asp 142
Glu 144
sp 316
rp 220
Argadin

◄ **Fig. 15** Schematic drawings of protein–ligand interactions (*left*) and surface plots of the active site (*right*), for complexes of *Serratia marcescens* ChiB with $GlcNAc_5$ [167] and with argifin and argadin [166]. In the schematic drawings, only intermolecular hydrogen bonds are shown. In the surface representations, the protein surface is colored *gray*, with the exception of Trp-97 and Trp-220 (*blue*) and the catalytic acid Glu-144 (*red*). The ligands are shown in a stick representation with carbons colored *green*. The sugar subsites are labeled in the $GlcNAc_5$ complex, and only the central three sugars are shown. Reproduced with permission from [166]. © 2002 by the National Academy of Sciences of the USA

8.3 Argadin–ChiB Complex

Argadin has a similar "tail" to that of argifin, with an *N*-acetyl group replacing the *N*-methylcarbamoyl group. In the argifin complex, this tail serves to mimic some key interactions of the – 1 GlcNAc residue. However, argadin binds in a completely different orientation, with the tail near the top of the binding site, illustrating a completely different mode of structural mimicry (Figs. 14, 15). The side chain of His 1 extends into the pocket normally occupied by the *N*-acetyl group of GlcNAc, and its NE2 donates a hydrogen bond to Asp 142, which is in the "inactive" conformation, hydrogen bonding to Asp 140 instead of Glu 144. The His ND2 hydrogen bonds to Tyr 214, replacing an interaction of the *N*-methylcarbamoyl carbonyl oxygen. Also unlike argifin, argadin extends into the – 2 subsite, where D-Pro 4 makes hydrophobic contacts with Tyr 292. Like argifin, argadin also contacts Asp 316 at the top of the site, with the Asp side chain in the open position; the roof cannot be closed completely due to the presence of the peptide. In the + 1 and + 2 subsites, the backbone of the cyclic ring of argadin makes extensive hydrophobic contacts with the Trp 97 and Trp 220 residues on each side. The backbone O of Arg 2 accepts a hydrogen bond from Arg 294, mimicking the interaction of the GlcNAc + 1 NAc O.

The hydrogen bond from the *N*-acetyl amide nitrogen to Glu 144 is not mimicked by argadin, but His 5 O accepts an H-bond from the other side chain oxygen of this residue, similar to the interaction with GlcNAc O4.

The L-aminoadipic acid 2 side chain fits into subsite + 1 in roughly the same way as one side of the GlcNAc + 1 pyranosyl ring. Its carboxylate group does not hydrogen bond to the protein, but rather to the arginyl tail; these two long side chains extend into subsite + 2.

Argadin was found to be a ~1000-fold more potent inhibitor than argifin (K_i values of 20 nM and 33 μM, respectively). While each ligand binds to a similar area, neither covers as much area as the GlcNAc pentamer (neither ligand occupies all five subsites). The difference in inhibition constants was attributed to the fact that argadin binds more deeply within the site, and displaces more ordered water molecules, than argifin. Also, argadin presumably has fewer unfavorable interactions with the Asp 316/Pro317 loop, as the loop is displaced significantly in the argifin complex; although in neither peptide complex is this loop able to "close" completely as in the carbohydrate complex.

That the cyclic peptides show interactions with the + 1 and + 2 subsites suggests an excellent starting point for design of inhibitors with selective inhibition of only certain chitinases. The interactions of argadin with W220 in the + 2 subsite, though appearing from the structure not to be extensive, were shown to be important as evidenced by a 250-fold reduction in inhibitory potency in the W220A mutant. Since W220 is only found in certain chitinases, these results could be very useful in guiding the rational design of inhibitors. Inhibitors based on carbohydrate analogues (e.g. allosamidin) were shown only to interact with the nonreducing (– 1 to – 3) subsites. In a similar way to the interactions with antibodies, the interactions of peptide mimotopes with enzymes may allow selectivity to be introduced which was not possible with the natural carbohydrate ligands.

Structure determinations of both peptides with two other family 18 chitinases, AfChiB1 of *Aspergillus fumigatus*, and human chitotriosidase (HCHT) provided further insight into this system [170]. Both inhibitors showed similar binding modes as with the *S. marcescens* enzyme; the inhibitors' orientations, and their interactions with key catalytic residues, were conserved. The other two enzymes lack the long loop (roof) at the top of the active site, which may be advantageous. Some small differences with large effects on binding affinity were observed. For example, in AfChiB1/argifin, in the – 1 subsite, Trp 384 (counterpart of Trp 403 in *S. marcescens* ChiB) approaches the *N*-methyl-carbamoylarginyl tail more closely, and this may allow more favorable van der Waals interactions. For the argadin complexes, the interactions with the + 1 and + 2 subsites are generally similar, but in AfChiB1, a Phe residue replaces Trp 220 in the + 2 subsite, leading to reduced hydrophobic interactions and π/π stacking, which is probably responsible for weaker binding affinity to this enzyme. Replacement of Phe 251 by Ala abrogated binding of both argadin and argifin, confirming the importance of this interaction.

Detailed mutagenesis studies showed that the interactions with the conserved catalytic residues Asp 175, Glu 177, Tyr 245, Asp 246, and Arg 301 in AfChiB1 were important for peptide binding, as evidenced by decreases in binding affinity to variants of the enzyme with each of these residues individually mutated. Decreases ranged from only 4-fold (Asp 175) to no detectable binding (Glu 177, Tyr 245, Asp 246). The effects of these mutations on carbohydrate binding were different: they caused decreases in k_{cat}, but did not affect K_m, suggesting that these hydrogen bonding interactions are not important for substrate binding, but only for catalysis [168].

Overall, these chitinase inhibitors provide an interesting example of peptide–carbohydrate mimicry. As described above for several other systems, mimicry consists of a mixture of structurally mimetic interactions and other favorable interactions with the receptor. Of the structurally mimetic interactions, the mimicry by argifin of the *N*-acetyl group, and part of the pyranosyl ring, of one substrate GlcNAc residue, is particularly striking. That argadin has a similar tail, yet binds in a completely different mode, is also an important result. This provides yet another illustration of the definition of molecular mimicry: it is with respect to interactions with a receptor, and in this case, both ligands can provide structural mimicry by interacting with a pocket of the enzyme meant to accommodate the GlcNAc *N*-acetyl group. The peptides and carbohydrates cannot be considered mimics of each other in the absence of the receptor.

9
Protein–Carbohydrate Mimicry

9.1
Amylases

Structural mimicry of carbohydrates by proteins has been observed in several protein–protein complexes, and many features are similar to those which so far seem characteristic of peptide–carbohydrate mimicry. An excellent example is provided by the crystal structure of porcine pancreatic α-amylase (PPA) in complex with α-amylase inhibitor 1 (α-AI1) [171]. The backbone atoms of a hairpin loop of the inhibitor mimic the position of a flexible loop of the enzyme in its active conformation, while side chains from two loops of the inhibitor are positioned in the active site, mimicking interactions of the substrate with the enzyme. Hydrophobic interactions provided by several residues from the inhibitor mimic those observed in the PPA–acarbose complex [172]. In addition, two tyrosine residues of α-AI1 are positioned to form hydrogen bonds with the catalytic aspartate and glutamate residues of the enzyme. These are similar to those observed in the PPA–acarbose complex, and presumably to those of the PPA–substrate complex. The hydroxyl group of one tyrosine residue has a similar position to that of the catalytic water molecule that normally hydrolyzes the starch substrate. The other tyrosine occupies the position of the acarviosine ring of acarbose, while its hydroxyl group forms a hydrogen bond to the nucleophilic aspartate residue. An unusual combination of backbone dihedral angles is required for the tyrosine to assume this position.

In the PPA–α-AI1 complex, a flexible loop of the enzyme, which would normally contact the substrate, is pushed outward to allow entry of the inhibitor, and one of the key aspartate residues is held in a conformation similar to that observed in the free enzyme. Therefore, some changes relative to the carbohydrate complex are required in order to accommodate the inhibitor. The structurally mimetic interactions within the catalytic site are supplemented by other specific protein–protein interactions, with a substantial buried surface area at the interface, involving 50 residues of the enzyme [171].

Structural mimicry of the substrate is also observed in the complex between PPA and the proteinaceous inhibitor tendamistat [173]. Again, the inhibitor extends into the catalytic site, but contacts only one of the catalytic residues, the glutamate residue. The sequence of three residues of the inhibitor, WRY (Trp-18 to Tyr-20) is positioned in the active site, with the arginine residue forming hydrogen bonds with the catalytic Glu-233. The WRY sequence and residues Thr-55, Gly-59 and Tyr-60 make hydrophobic contacts with the active site which are similar to those of acarbose and of α-AI1.

Contacts with the catalytic residues, in combination with hydrophobic interactions, are also observed in the complex of an insect α-amylase with the *Ragi* bifunctional α-amylase/trypsin inhibitor (RBI) [174]. Conversely, the mechanism of inhibition of barley α-amylase by the barley α-amylase/subtilisin inhibitor (BASI) did not involve direct contact between inhibitor residues and the catalytic site [175]. The inhibitor sterically blocks the catalytic site, but does not extend into it. A cavity is created, which is occupied by a calcium ion coordinated by water-mediated interactions with the catalytic residues.

9.2 Lysozyme

Another example of mimicry is provided by a camel heavy-chain antibody directed against the catalytic site of lysozyme. The crystal structure of this complex was compared with crystal structures of lysozyme in complex with carbohydrates [176, 177]. Almost all residues of lysozyme contacted by the substrate are also contacted by the antibody, and several important interactions are mimicked. As in the SYA/J6 complex, an interaction between the *N*-acetyl methyl group of GlcNAc and a pocket formed by hydrophobic side chains is observed. The side chain of Ala-104 in the camel Ab mimics this methyl group, and is positioned in the same hydrophobic pocket (Fig. 16). Several hydrogen bonds are mimicked as well: for example, the amide nitrogen and oxygen of Ala-104 adopt similar positions to those of the GlcNAc acetamido group of the substrate, and form hydrogen bonds to the same lysozyme atoms. The hydrogen bonding of substrate hydroxyl groups is performed by carbonyl or hydroxyl groups of the camel Ab; in total, 5 of 12 hydrogen bonds in the carbohydrate–lysozyme complex are mimicked by the Ab. Several areas of hydrophobic contact are also similar [177].

9.3 Anti-Idiotopic Antibodies

Idiotope refers to the binding site of an antibody and also implies the antibody's specificity. In chemical terms, the idiotope implies a particular binding site with a certain sequence and structure. Antibodies directed against the

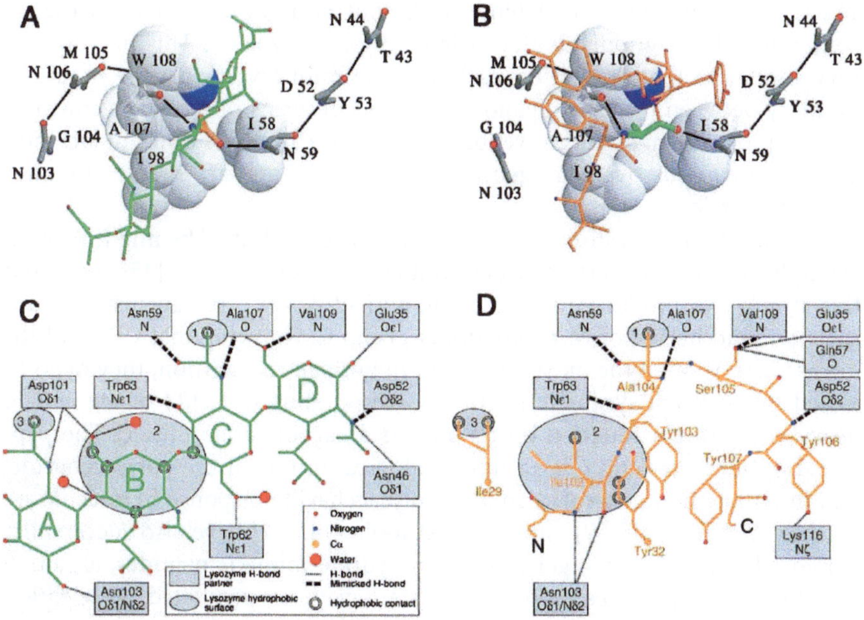

Fig. 16 Substrate mimicry in the lysozyme active site by a camel heavy-chain antibody. The hydrophobic pocket of lysozyme subsite C and its interactions with the carbohydrate substrate (**A**) and camel heavy-chain antibody (**B**). The lysozyme residues Ile-58, Ile-98, Ala-107, and Trp-108 are shown as *gray*, space-filling models, with Cα atoms labeled. For Ala-107, Cβ is labeled and is *semi-transparent*. The *N*-acetamido methyl group of the carbohydrate (**A**) and the Ala-104 side chain of the inhibitor (**B**) are observed in very similar positions in the pocket. In addition, the backbone HN and O atoms of Ala-104 form similar hydrogen bonds as the HN and O atoms of the *N*-acetamido group of the carbohydrate. **C** and **D** show schematic diagrams of hydrogen-bonding and hydrophobic interactions within the lysozyme–carbohydrate (**C**) and lysozyme–protein (**D**) complexes. *Gray boxes* indicate lysozyme atoms involved in hydrogen bonds. Hydrogen bonds unique to each ligand are shown as *thin dashed lines*, while hydrogen bonds mimicked by the protein ligand are shown as *bold dashed lines*. Three hydrophobic areas on lysozyme are shown as *gray ovals*: 1, the side chains of Ile-58, Ile-98, Ala-107, and Trp-108, corresponding to the pocket shown in (**A**) and (**B**); 2, the side chain of Trp-62; and 3, the side chains of Leu75 and Trp62. Hydrophobic interactions that are similar for both the carbohydrate and protein, with these areas of lysozyme, are indicated by double circles around the atoms of the ligands. Reproduced with permission from [177]. © 1998 Wiley-Liss, Inc.

binding site of other Abs are known as anti-idiotopic, and are thought to exist physiologically and play a role in regulation of the immune system. Jerne presented the hypothesis that some anti-idiotopic Abs might structurally resemble the antigen, or bear its "internal image" [178]; in other words, would provide structural mimicry of the antigen. Anti-idiotopes have been classified as those with structural mimicry of the antigen (type β); those that recognize the binding site but do not provide structural mimicry of the antigen (type γ) and those that bind an Ab but at a site distant from the Ab combining site (e.g. outside the variable region) (type α). This paradigm is very similar to that of peptide–carbohydrate mimicry, in that the nature of mimicry between different ligands is defined in terms of the involvement of the receptor. The main incentive for studying anti-Id Abs is also similar: their possible use as surrogate antigens, with potentially more favorable immunogenic properties.

Anti-idiotopic Abs directed against anti-carbohydrate Abs might provide examples of protein–carbohydrate mimicry. Westerink et al. [45] employed this strategy to produce a monoclonal, anti-Id Ab directed against the capsular polysaccharide of *N. meningitidis* group C. Knowing that CDR3 of the heavy chain is often the most important in antigen recognition, they hypothesized that a small synthetic peptide corresponding to CDR3 of the anti-Id heavy chain might function on its own as a carbohydrate mimic. The peptide did indeed function as an immunological mimic of the polysaccharide, and immunization generated an anti-polysaccharide response in mice. More recently, Harris et al. [59] and Ngamastupol et al. [77] have also successfully employed this strategy to generate carbohydrate-mimetic peptides, although the majority of peptides reported to date have been generated by selection from phage-display libraries.

Evans et al. [179] produced anti-idiotopic Abs directed against the mAb YsT9.1, which recognizes the lipopolysaccharide of type A *Brucella abortus*, an α-(1-2)-linked polymer of 4,6-dideoxy-4-formamido-α-D-mannopyranose. These Abs competed with the polysaccharide for binding to YsT9.1, but were unable to induce an anti-polysaccharide Ab response. The structure of the complex of the monoclonal anti-Id Ab (Ab2) with mAb YsT9.1 (Ab1) revealed the reasons for this: the binding site of Ab1 is a long, narrow groove, 10 Å deep, and the binding site of Ab2, though it has a convex shape and extends to contact the top and sides of the Ab1 binding site, does not extend to the bottom of the site, and does not provide complete structural mimicry of the polysaccharide. Ab2 recognition of Ab1 also involved "induced fit": the bound conformation of Ab2 differed from the unliganded conformation, especially in the region of CDR H2, in which a rearrangement of an Asp side chain by 5 Å was necessary to create a complementary conformation. This rearrangement was likely to be unfavorable, as it involved burial of the Asp side chain in a hydrophobic pocket without hydrogen bonding or electrostatic interactions. This Gly-rich loop of Ab2 may be somewhat flexible in solution, with only a minor population of the conformation that is complementary to Ab1, which would be another possible reason why Ab2 was unable to function as an immunogenic mimic of the polysaccharide.

Further studies showed that the frequency of occurrence of a polysaccharide-mimicking idiotope was low [180]. Immunizations with three different anti-Id Abs failed to produce a polysaccharide-reactive response or produced only a weak response, implying that structural mimicry of a polysaccharide is difficult to achieve with the Ab repertoire. In some cases, anti-polysaccharide Abs have been produced successfully by anti-Id immunization [32, 181–183]; however, to date structural information is not available for these complexes.

Structural information is available for several anti-Id Abs mimicking protein antigens [184]. The structures of complexes of an anti-lysozyme antibody with lysozyme and with an anti-Id Ab showed a certain amount of structural mimicry, supplemented by other interactions. Roughly half of the hydrogen bonds in the lysozyme/Ab interface were mimicked by the anti-Id Ab, and 78% of the interface was conserved in terms of hydrophilicity/hydrophobicity, indicating interactions which would be similar if not identical. It was proposed that a certain proportion of identical interactions with the Ab1 binding site must be made by the Ag and by Ab2 in order to provide immunologically functional mimicry. The anti-Id Ab also acted as a functional immunogenic mimic in eliciting anti-lysozyme Abs, but the anti-lysozyme response was only a small fraction of the total anti-Ab response.

10 Glycopeptides

In a strategy largely separate from the studies described above of peptide–carbohydrate mimicry, several groups have investigated glycopeptides as mimetic ligands for carbohydrate-binding proteins (reviewed in [185]). Meldal et al. designed glycopeptides with peptidic scaffolds designed to present saccharide residues in the correct orientation [186], and investigated possible structural similarity between the glycopeptides and oligosaccharides by molecular dynamics simulations [187]. Although in these cases the peptidic portion was designed as a scaffold, the hypothesis that the peptidic portion might also interact with the binding site and contribute binding energy was also presented [185–187].

Lin et al. created a series of fucopeptides as mimics of sialyl Lewis X, with a fucose residue linked to a variety of peptidic scaffolds, and identified several compounds which functioned as effective inhibitors of E-selectin with greater binding affinity than SLe^X [188]. Several groups have designed glycopeptide libraries, using different synthetic strategies to create combinatorial libraries with variability in both the peptide and carbohydrate portions [185]. Recently, Xian et al. described libraries of C-linked glycopeptides designed to mimic the Gal-α-(1-3)-Gal epitope. Considering the importance of the terminal nonreducing Gal residue, this residue was maintained while randomized peptide sequences were linked to it in an α configuration. Several new ligands were identified and shown to have comparable affinity as the trisaccharide Gal-α-(1-3)-Gal-β-(1-4)-GlcNAc in binding to anti-Gal antibodies [189]. Ying et al. created libraries of glycosylated peptides, in which an *N*-glycosylated Asn residue formed part of a randomized peptide sequence, and identified several glycopeptide ligands for wheat germ agglutinin (WGA). Molecular modeling studies using the crystal structure of WGA supported possible binding modes similar to known WGA-oligosaccharide complexes, in which the GlcNAc residue maintained known interactions while the peptide portions contributed other interactions. These interactions varied depending on the peptide sequence, but often included hydrophobic interactions with a Tyr residue, similar to those of saccharides, and hydrogen bonds to various other residues in the binding site [190]. Debenham et al. synthesized C-linked glycopeptides that bound in a bivalent manner to concanavalin A [191]. Halkes et al. designed sialopeptide libraries including O-linked NeuAc-Thr and identified high-affinity ligands for sialoadhesin, and again, molecular modeling supported a binding mode in which sialic acid adopts a known position and certain amino acid residues create other favorable interactions. The modeling study was consistent with the consensus sequence isolated from the library, in that the invariant residues were found to provide multiple interactions while the variable *N*-terminal peptide portion made few contacts [192].

In a strategy employing rational design rather than combinatorial selection, glycopeptides combining parts of peptide and carbohydrate ligands for an anti-*Shigella* antibody were created (see Sect. 3) [193]. Since the peptide and carbohydrate ligands interacted with the antibody using quite different sets of interactions [80], detailed molecular modeling studies were required to identify possible glycopeptide ligands. A compound with an *N*-terminal tripeptide portion of the carbohydrate-mimetic peptide, linked via a thioether linkage to a trisaccharide Rha-α-(1-2)-Rha-α-(1-3)-Rha, representing part of the *Shigella flexneri* type Y lipopolysaccharide, was identified as an optimal second-generation ligand. Detailed studies of the binding of this ligand are in progress.

Glycopeptides present a promising strategy to combine the advantages of carbohydrate and peptide ligands. Here, the basis of mimicry is predicted to be similar to the above-described peptide–carbohydrate mimicry, and it is postulated that "structurally mimetic" interactions would be provided by the carbohydrate portion—typically one sugar residue—and in some cases by the peptide portion, which might provide similar interactions as the missing oligosaccharide residues. As in the case of peptide–carbohydrate mimicry, additional favorable interactions not representative of structural mimicry are expected to be provided by the peptide component. It should be noted that to date, structural studies of these complexes are not available. Most molecular modeling studies have assumed that the binding mode of the sugar residue, in most cases known to be important for molecular recognition, would be similar to that in the oligosaccharide complex, and in some cases the sugar residue has been restrained in this position, in order to make computations more tractable. Structural data on these complexes will presumably reveal whether this is the case, and if so, whether some small adjustments of the binding mode may occur; very small adjustments have been correlated with major differences in binding affinity [87, 88]. Alternatively, completely different binding modes might be adopted, as has been observed in the above-described case of argifin and argadin, and in several other cases involving similar ligands.

11
New Applications and Challenges for the Therapeutic Use of Carbohydrate-Mimetic Peptides

As described above, peptides may be found which act as ligands for various carbohydrate-binding proteins. In the case of enzymes, it is sufficient to identify compounds which compete for binding with the carbohydrate substrate, and which inhibit enzyme action. To establish specificity still remains a challenge. For example, the chitinase inhibitors described above are potential agents against pathogenic fungi, such as *Aspergillus*, which cause infections in immunocompromised individuals. Discrimination between human chitinase enzymes and the microbial enzymes would be necessary to avoid toxic side effects, involving a significant difference in K_D. The structures suggest several possible ways that the differences in interaction patterns between the peptides and the carbohydrate substrate could be exploited to achieve this.

The challenges are similar for the case of lectins. Peptide ligands for E-selectin are important lead compounds in the development of anti-inflammatory drugs [18, 33, 34]. For example, peptides discovered by screening phage-display libraries with E-selectin (e.g. DITWDQLWDLMK, IC_{50} 4.0 nM; sialyl Le^X, IC_{50} 750 μM), inhibited leukocyte rolling and adhesion to E-selectin, and diminished acute inflammation reactions in mice [18]. However, these peptides were not structural mimics of sialyl Le^X, as they did not compete with the carbohydrate for binding to E-selectin. In contrast, the peptide IELLQAR, selected for its binding to an anti-Le^A antibody, did compete with sialyl Le^X for binding to E-selectin (IC_{50} 200 μM), and also bound to L- and P-selectins. The peptide inhibited cell adhesion to E-selectin and metastasis of tumor cells [34].

Applications in neurology may also be possible. A peptide selected by a mAb directed against the L2/HNK-1 epitope, the trisaccharide GlcA-5-sulfate-β(1-3)-Gal-β-(1-4)-GlcNAc, was not recognized by the HNK-1 mAb, which requires the sulfate group for binding, but was recognized by another mAb which reacts with both sulfated and nonsulfated forms [38]. The peptide was shown to compete for binding to glycolipids on a solid phase, to bind to motor neurons and to enhance axon/neurite outgrowth and polarity. In another example, two cyclic peptides were identified as mimics of poly-α-(2-8)-NeuAc. These peptides were specific for their isolating mAb and had effects on neuronal cells in vivo which were poly-sialic acid dependent, possibly involving effects on an unknown receptor. These functional mimics have potential applications in the treatment of neurodegenerative diseases and of central nervous system injury.

The use of carbohydrate-mimetic peptides as vaccines presents greater challenges. Rather than considering the interaction of the ligand with one receptor, interactions with the entire Ab repertoire present in vivo, as well as the effects on B and T cells and the cytokine signals that control development of the immune response, must be investigated. Thus, unlike the case of enzyme inhibition, competition for binding to the receptor of interest, usually a monoclonal antibody, is not sufficient for biological activity. The peptide must also be able to induce an anti-carbohydrate immune response. The peptide must not only bind to the isolating mAb, but must bind to a similar, natural Ab displayed on the surface of B cells in vivo; binding must be effective enough to stimulate this B cell clone to expand and overproduce the desired Ab, and in the ideal case, to stimulate a memory response, affinity maturation and isotype switching (characteristics not usually induced by carbohydrate Abs, as they require T cell help for which antigen presentation on MHC molecules is essential).

As is documented in Table 2, the induction of anti-carbohydrate immune responses has been successfully demonstrated for many peptides. However, certain challenges remain. Often, the induced antibody titer (concentration) directed against the carbohydrate is rather weak, especially in comparison to the anti-peptide response; the majority of antibodies recognize the peptide presumably in a nonmimetic conformation. Several recent studies have investigated these immune responses in more detail, with an extensive analysis of antibody isotypes and adjuvant effects [70, 75], the effect of different protein carriers [194], the use of DNA vaccines [48, 195], and the demonstration of protection by passive immunization [72, 75, 194]. In several cases, the anti-carbohydrate immune response has also been shown to be protective against infection in mice [30, 72, 75, 78], and these cases are probably the most promising for the development of vaccines.

Several studies have shown that peptides can discriminate between closely related mAbs [20, 25, 140, 196], raising the possibility of selecting only desired components of an immune response. This may be particularly important in situations such as *Cryptococcus neoformans*, where antibodies are not only protective but also nonprotective and even disease-enhancing. However, peptides are not necessarily more specific in all cases: a recent study showed that a peptide mimic selected by an Ab directed against *Streptococcus pneumoniae* type 6B capsular polysaccharide was identical to that selected by an Ab against *N. meningitidis* group B capsular polysaccharide; the Abs themselves were not related and their V region sequences were quite different [36]. There is therefore still a possibility of unintended immune responses against other antigens. The best possibility for the application of peptide mimotopes will likely be as part of a prime/boost strategy involving immunization with both carbohydrate and peptide antigens.

The use of peptides as anticancer vaccines is also a promising area. Peptide mimics of the GD1α ganglioside [28] and the GD2 ganglioside [76] were shown to inhibit tumor growth and metastasis. Peptides isolated for their affinity to anti-Lewis Y mAbs (see Sect. 7.2), or to lectins directed against cell-surface carbohydrates, also have applications as anticancer vaccines ([56]; reviewed in [197–199]). Similar strategies, exploiting the functional mimicry between peptides and α-mannopyranosides (see Sects. 6 and 7.1) are being employed in the development of peptide mimotopes as anti-HIV vaccines [64, 74]. Here, the basis of mimicry is still unclear, as in some cases the isolating receptor is a lectin, while the functionally relevant receptors are anti-HIV Abs, and the mechanism of mimicry likely differs between each receptor.

12 Conclusions

Some general features of peptide–carbohydrate mimicry have become clear from recent studies, while others remain to be clarified. Peptide ligands have been identified for many carbohydrate-binding proteins, including antibodies, lectins, and enzymes. In most cases, the selected peptide is specific for the protein receptor, rather than for the carbohydrate it "mimics". Thus, if different antibodies directed against the same carbohydrate targets are screened, different peptides are usually selected. A corollary is that conclusions regarding structural mimicry cannot be made by superposition of the ligands. Mimicry is only meaningful in the context of complementary interactions with proteins.

The binding affinity of carbohydrate-mimetic peptides can be quite high, often higher than the natural carbohydrate ligand. Carbohydrate-mimetic peptides do not share any general chemical features, and all amino acid residues are represented; there is no evidence for an unusual importance of aromatic residues in mimicry. In several cases, D-amino acids have been incorporated, and unusual amino acids (derived from nonribosomal peptide synthesis) have been found in naturally occurring carbohydrate-mimetic peptides.

NMR spectroscopy and X-ray crystallography have been applied to investigate the bioactive conformations and epitopes of carbohydrate and mimetic peptide ligands. The bound conformations of peptide mimetics for *Cryptococcus neoformans* and for groups A and B *Streptococcus* all include tight turns; these conformations likely provide some intramolecular stabilization as well as exposing peptide side chains for intermolecular interactions. In general, peptides are capable of adopting multiple conformations of similar energy, and therefore, peptides with a significant preference for the bioactive turn conformation may be the most effective. This was demonstrated for peptide mimotopes of the group B *Streptococcus* capsular polysaccharide and of the group A *Streptococcus* cell-wall polysaccharide, which were shown to populate their bioactive conformations in the absence of antibody, and to be effective immunogenic mimics.

Complete comparisons of the structures of protein receptors with bound peptides and carbohydrates have been possible in only two cases to date: the anti-*Shigella* antibody SYA/J6 and the family 18 chitinases. Peptide–carbohydrate mimicry appears to consist of a minority of structurally mimetic interactions, meaning intermolecular interactions (hydrogen bonds or van der Waals contacts) which are similar between the mimetic ligand and natural ligand; and a greater number of other favorable interactions. It appears that peptides cannot exactly mimic the shapes of the bound conformations of carbohydrates, but can nevertheless provide topographies complementary to the protein receptor, often extending into different areas of the binding site which are not contacted by the carbohydrate. A certain small fraction of structural mimicry is usually present; for example, mimicry of GlcNAc *N*-acetamido methyl groups by peptide alanine side chains has been observed in several cases. All of these observations also hold for the computationally predicted complexes of carbohydrates and peptides in several systems, and for several examples of protein–carbohydrate mimicry.

Acknowledgements Financial support from the Natural Sciences and Engineering Research Council of Canada, the Canadian Institutes of Health Research (fellowship to M.A.J.) and the Skaggs Institute for Chemical Biology is gratefully acknowledged. This is manuscript #17973-MB from the Scripps Research Institute.

References

1. Cunningham MW, Fujinami RS (eds) (2000) Molecular mimicry, microbes, and autoimmunity. ASM Press, Washington, DC
2. Oldstone MBA (1998) FASEB J 12:1255
3. Elbein AD, Molyneux RJ (1999) Alkaloid glycosidase inhibitors. In: Pinto BM (ed) Comprehensive natural products chemistry, vol 3, chap 7. Elsevier, London
4. Stütz AE (ed) (1999) Iminosugars as glycosidase inhibitors: nojirimycin and beyond. Wiley-VCH, New York
5. Jennings HJ (1990) Curr Top Microbiol Immunol C125:373
6. Jennings HJ (1983) Adv Carbohydr Chem Biochem 41:155
7. Hakomori S (1996) Cancer Res 56:5309
8. Dube D, Bertozzi C (2005) Nat Rev Drug Discov 4:477
9. Beenhouwer DO, May R, Valadon P, Scharff M (2002) J Immunol 169:6992
10. Oldenburg KR, Loganathan D, Goldstein IJ, Schultz PG, Gallop MA (1992) Proc Nat Acad Sci USA 89:5393

11. Scott JK, Loganathan D, Easley RB, Gong X, Goldstein IJ (1992) Proc Nat Acad Sci USA 89:5398
12. Hoess R, Brinkmann U, Handel T, Pastan I (1993) Gene 128:43
13. Johnson MA, Pinto BM (2002) Aust J Chem 55:13
14. Bonnycastle LL, Mehroke JS, Rashed M, Gong X, Scott JK (1996) J Mol Biol 258:747
15. Shikhman AR, Cunningham MW (1994) J Immunol 152:4375
16. Ravindranath RMH, Tam WY, Nguyen P, Fincham AG (2000) J Biol Chem 275:39654
17. Bianchi E, Folgori A, Wallace A, Nicotra M, Acali S, Phalipon A, Barbato G, Bazzo R, Cortese R, Felici F, Pessi A (1995) J Mol Biol 247:154
18. Martens CL, Cwirla SE, Lee RYW, Whitehorn E, Chen EYF, Bakker A, Martin EL, Wagstrom C, Gopalan P, Smith CW, Tate E, Koller KJ, Schatz PJ, Dower WJ, Barrett RW (1995) J Biol Chem 270:21129
19. Eichler J, Lucka AW, Pinilla C, Houghten RA (1996) Mol Diversity 1:233
20. Valadon P, Nussbaum G, Boyd LF, Margulies DH, Scharff MD (1996) J Mol Biol 261:11
21. Valadon P, Nussbaum G, Oh J, Scharff MD (1998) J Immunol 161:1829
22. Young ACM, Valadon P, Casadevall A, Scharff MD, Sacchettini JC (1997) J Mol Biol 274:622
23. Taki T, Ishikawa D, Hamasaki H, Handa S (1997) FEBS Lett 418:219
24. Nussbaum G, Cleare W, Casadevall A, Scharff MD, Valadon P (1997) J Exp Med 185:685
25. Harris SL, Craig L, Mehroke JS, Rashed M, Zwick MB, Kenar K, Toone EJ, Greenspan N, Auzanneau FI, Marino-Albernas JR, Pinto BM, Scott JK (1997) Proc Nat Acad Sci USA 94:2454
26. Apostolopoulos V, Sandrin MS, McKenzie IFC (1999) J Mol Med 77:427
27. Apostolopoulos V, Lofthouse SA, Popovski V, Chelvanayagam G, Sandrin MS, McKenzie IFC (1998) Nat Biotechnol 16:276
28. Ishikawa D, Kikkawa H, Ogino K, Hirabayashi Y, Oku N, Taki T (1998) FEBS Lett 441:20
29. Pinilla C, Appel JR, Campbell GD, Buencamino J, Benkirane N, Muller S, Greenspan NS (1998) J Mol Biol 283:1013
30. Pincus SH, Smith MJ, Jennings HJ, Burritt JB, Glee PM (1998) J Immunol 160:293
31. Moe GR, Tan S, Granoff DM (1999) FEMS Immunol Med Microbiol 26:209
32. De Bolle X, Laurent T, Tibor A, Godfroid F, Weynants V, Letesson JJ, Mertens P (1999) J Mol Biol 294:181
33. O I, Kieber-Emmons T, Otvos L, Blaszczyk-Thurin M (2000) Biochem Biophys Res Commun 268:106
34. Fukuda MN, Ohyama C, Lowitz K, Matsuo O, Pasqualini R, Ruoslahti E, Fukuda M (2000) Cancer Res 60:450
35. Shin JS, Lin JS, Anderson PW, Insel RA, Nahm MH (2001) Infect Immun 69:3335
36. Ziebell MR, Zhao ZG, Luo B, Luo Y, Turley EA, Prestwich GD (2001) Chem Biol 8:1081
37. Shin JS, Yu J, Lin J, Zhong L, Bren KL, Nahm MH (2002) J Immunol 168:6273
38. Simon-Haldi M, Mantei N, Franke J, Voshol H, Schachner M (2002) J Neurochem 83:1380
39. Cosman M, Lightstone FC, Krishnan VV, Zeller L, Prieto MC, Roe DC, Balhorn R (2002) Chem Res Toxicol 15:1218
40. Arnusch CJ, Andre S, Valentini P, Lensch M, Russwurm R, Siebert HC, Fischer MJE, Gabius HJ, Pieters RJ (2004) Bioorg Med Chem Lett 14:1437

41. Lauvrak V, Berntzen G, Heggelund U, Herstad TK, Sandin RH, Dalseg R, Rosenqvist E, Sandlie I, Michaelsen TE (2004) Scand J Immunol 59:373
42. Torregrossa P, Buhl L, Bancila M, Durbec P, Schafer C, Schachner M, Rougon G (2004) J Biol Chem 279:30707
43. Dinglasan RR, Porter-Kelley JM, Alam U, Azad AF (2005) Vaccine 23:2717
44. Förster-Waldl E, Riemer AB, Dehof AK, Neumann D, Brämswig K, Boltz-Nitulescu G, Pehamberger H, Zielinski CC, Scheiner O, Pollak A, Lode H, Jensen-Jarolim E (2005) Mol Immunol 42:319
45. Westerink MAJ, Giardina PC, Apicella MA, Kieber-Emmons T (1995) Proc Nat Acad Sci USA 92:4021
46. Hutchins WA, Kieber-Emmons T, Carlone GM, Westerink MA (1999) Hybridoma 18:121
47. Phalipon A, Folgori A, Arondel J, Sgaramella G, Fortugno P, Cortese R, Sansonetti PJ, Felici F (1997) Eur J Immunol 27:2620
48. Prinz DM, Smithson SL, Kieber-Emmons T, Westerink MAJ (2003) Immunol 110:242
49. Luo P, Agadjanyan M, Qiu J, Westerink MAJ, Steplewski Z, Kieber-Emmons T (1998) Mol Immunol 35:865
50. Agadjanyan M, Luo P, Westerink MAJ, Carey LA, Hutchins W, Steplewski Z, Weiner DB, Kieber-Emmons T (1997) Nat Biotechnol 15:547
51. Cunto-Amesty G, Dam TK, Luo P, Monzavi-Karbassi B, Brewer CF, Van Cott TC, Kieber-Emmons T (2001) J Biol Chem 276:30490
52. Kaur KJ, Khurana S, Salunke DM (1997) J Biol Chem 272:5539
53. Kaur KJ, Jain D, Goel M, Salunke DM (2001) Vaccine 19:3124
54. Lou Q, Pastan I (1999) J Peptide Res 53:252
55. Luo P, Canziani G, Cunto-Amesty G, Kieber-Emmons T (2000) J Biol Chem 275:16146
56. Monzavi-Karbassi B, Artaud C, Jousheghany F, Hennings L, Carcel-Trullols JC, Shaaf S, Korourian S, Kieber-Emmons T (2005) J Immunol 174:7057
57. Grothaus MC, Srivastava N, Smithson SL, Kieber-Emmons T, Williams DB, Carlone GM, Westerink MA (2000) Vaccine 18:1253
58. Charalambous BM, Feavers IM (2000) FEMS Microbiol Lett 191:45
59. Harris SL, Park MK, Nahm MH, Diamond B (2000) Infect Immun 68:5778
60. Fleuridor R, Lees A, Pirofski L (2001) J Immunol 166:1087
61. Zhang H, Zhong Z, Pirofski LA (1997) Infect Immun 65:1158
62. Lesinski GB, Smithson SL, Srivastava N, Chen D, Widera G, Westerink MAJ (2001) Vaccine 19:1717
63. Jouault T, Fradin C, Dzierszinski F, Borg-Von-Zepelin M, Tomavo S, Corman R, Trinel PA, Kerckaert JP, Poulain D (2001) Glycobiology 11:693
64. Pashov A, Canziani G, Monzavi-Karbassi B, Kaveri SV, MacLeod S, Saha R, Perry M, Van Cott TC, Kieber-Emmons T (2005) J Biol Chem 280:28959
65. Gulati S, Ngampasutadol J, Yamasaki R, McQuillen DP, Rice PA (2001) Int Rev Immunol 20:229
66. Melzer H, Fortugno P, Mansouri E, Felici F, Marinets A, Wiedermann G, Kollaritsch H, von Specht BU, Duchêne M (2002) Parasite Immunol 24:321
67. Melzer H, Baier K, Felici F, von Specht BU, Wiedermann G, Kollaritsch H, Wiedermann U, Duchêne M (2003) FEMS Immunol Med Microbiol 37:179
68. Harris SL, Dagtas AS, Diamond B (2002) Mol Immunol 39:263
69. Brett PJ, Tiwana H, Feavers IM, Charalambous BM (2002) J Biol Chem 277:20468
70. Hou Y, Gu XX (2003) J Immunol 170:4373
71. Prinz DM, Smithson SL, Westerink MAJ (2004) J Immunol Methods 285:1

72. Park I, Choi IH, Kim SJ, Shin JS (2004) Yonsei Med J 45:755
73. Buchwald UK, Lees A, Steinitz M, Pirofski LA (2005) Infect Immun 73:325
74. Pashov A, Canziani G, MacLeod S, Plaxco J, Monzavi-Karbassi B, Kieber-Emmons T (2005) Vaccine 23:2168
75. Gevorkian G, Segura E, Acero G, Palma JP, Espitia C, Manoutcharian K, López-Marín LM (2005) Biochem J 387:411
76. Bolesta E, Kowalczyk A, Wierzbicki A, Rotkiewica P, Bambach B, Tsao CY, Horwacik I, Kolinski A, Rokita H, Brecher M, Wang X, Ferrone S, Kozbor D (2005) Cancer Res 65:3410
77. Ngampasutadol J, Rice PA, Walsh MT, Gulati S (2006) Vaccine 24:157
78. Falklind-Jerkerus S, Felici F, Cavalieri C, Lo Passo C, Garufi G, Pernice I, Islam MM, Qadri F, Weintraub A (2005) Microbes Infect 7:1453
79. Pitner JB, Beyer WF, Venetta TM, Nycz C, Mitchell MJ, Harris SL, Mariño-Albernas JR, Auzanneau FI, Forooghian F, Pinto BM (2000) Carbohydr Res 324:17
80. Vyas NK, Vyas MN, Chervenak MC, Bundle DR, Pinto BM, Quiocho FA (2003) Proc Nat Acad Sci USA 100:15023
81. Vyas MN, Vyas NK, Meikle PJ, Sinnott B, Pinto BM, Bundle DR, Quiocho FA (1993) J Mol Biol 231:133
82. Vyas NK, Vyas MN, Chervenak MC, Johnson MA, Pinto BM, Bundle DR, Quiocho FA (2002) Biochemistry 41:13575
83. Carlin NIA, Gidney MAJ, Lindberg AA, Bundle DR (1986) J Immunol 137:2361
84. Carlin NIA, Lindberg AA, Bock K, Bundle DR (1984) Eur J Biochem 139:189
85. Wu TT, Kabat EA (1970) J Exp Med 132:211
86. Al-Lazikani B, Lesk AM, Chothia C (1997) J Mol Biol 273:927
87. Bundle DR (1998) In: Hecht S (ed) Bioorganic chemistry: carbohydrates. Oxford University Press, Oxford, pp 370–439
88. Bundle DR (1989) Pure Appl Chem 61:1171
89. Jeffrey PD, Bajorath J, Chang CY, Yelton D, Hellström I, Hellström KE, Sheriff S (1995) Nat Struct Biol 2:466
90. Cygler M, Rose DR, Bundle DR (1991) Science 253:442
91. Herron JN, He XM, Mason ML, Voss EW, Edmundson AB (1989) Proteins 5:271
92. Wilson IA, Stanfield RL (1994) Curr Opin Struct Biol 4:857
93. Braden BC, Poljak RJ (1995) FASEB J 9:9
94. Davies DR, Cohen GH (1996) Proc Nat Acad Sci USA 93:7
95. Mariuzza RA, Poljak RJ (1993) Curr Opin Immunol 5:50
96. Mayer M, Meyer B (1999) Angew Chem Int Ed 38:1784
97. Mayer M, Meyer B (2001) J Am Chem Soc 123:6108
98. Meyer B, Peters T (2003) Angew Chem Int Ed 42:864
99. Johnson MA, Pinto BM (2004) Carbohydr Res 339:907
100. Johnson MA, Pinto BM (2004) Bioorg Med Chem 12:295
101. Jayalakshmi V, Krishna NR (2002) J Magn Reson 155:106
102. Harris SL (1997) PhD thesis, Simon Fraser University
103. Johnson MA, Eniade AA, Pinto BM (2003) Bioorg Med Chem 11:781
104. Hossany RB, Johnson MA, Eniade AA, Pinto BM (2004) Bioorg Med Chem 12:3743
105. Baker CJ, Edwards MS (1988) Ann NY Acad Sci 549:193
106. Zou W, Mackenzie R, Thérien L, Hirama T, Yang Q, Gidney MA, Jennings HJ (1999) J Immunol 163:820
107. Brisson JR, Uhrinova S, Woods RJ, van der Zwan M, Jarrell HC, Paoletti LC, Kasper DL, Jennings HJ (1997) Biochemistry 36:3278

108. Johnson MA, Jaseja M, Zou W, Jennings HJ, Copie V, Pinto BM, Pincus SH (2003) J Biol Chem 278:24740
109. Brisson JR, Baumann H, Imberty A, Pérez S, Jennings HJ (1992) Biochemistry 31:4996
110. Evans SV, Sigurskjold BW, Jennings HJ, Brisson JR, To R, Tse WC, Altman E, Frosch M, Weisgerber C, Kratzin HD, Klebert S, Vaesen M, Bitter-Suermann D, Rose DR, Young NM, Bundle DR (1995) Biochemistry 34:6737
111. Michon F, Moore SL, Kim J, Blake MS, Auzanneau FI, Johnston BD, Johnson MA, Pinto BM (2005) Infect Immun 73:6383
112. Pinto BM (1992) Synthesis and immunochemistry of carbohydrate antigens of the β-hemolytic *Streptococcus* group A. In: Garegg PJ, Lindberg AA (eds) Carbohydrate antigens. ACS Symp Ser 519. American Chemical Society, Washington, pp 111–131
113. Marino-Albernas JR, Harris SL, Varma V, Pinto BM (1993) Carbohydr Res 245:245
114. Auzanneau FI, Forooghian F, Pinto BM (1996) Carbohydr Res 291:21
115. Reimer KB, Gidney MAJ, Bundle DR, Pinto BM (1992) Carbohydr Res 232:131
116. Auzanneau FI, Pinto BM (1996) Bioorg Med Chem 4:2003
117. Auzanneau FI, Christensen MK, Harris SL, Meldal M, Pinto BM (1998) Can J Chem 76:1109
118. Coligan JE, Kindt TJ, Krause RM (1978) Immunochem 15:755
119. Huang DH, Krishna NR, Pritchard DG (1986) Carbohydr Res 155:193
120. Kabat EA (1968) Structural concepts in immunology and immunochemistry. Holt, Rinehart and Winston, New York
121. Shikhman AR, Cunningham MW (1994) J Immunol 152:4375
122. Cunningham MW (2000) Molecular mimicry between streptococcal M protein and cardiac myosin and the immunopathogenesis of rheumatic fever. In: Cunningham MW, Fujinami RS (eds) Molecular mimicry, microbes, and autoimmunity. ASM Press, Washington, DC, pp 39–56
123. Weimar T, Harris SL, Pitner JB, Bock K, Pinto BM (1995) Biochemistry 34:13672
124. Johnson MA, Pinto BM (2002) J Am Chem Soc 124:15368
125. Johnson MA, Rotondo AA, Pinto BM (2002) Biochemistry 41:2149
126. Zwahlen C, Vincent SJF, Di Bari L, Levitt MH, Bodenhausen G (1994) J Am Chem Soc 116:362
127. Vincent SJF, Zwahlen C, Bodenhausen G (1996) J Biomol NMR 7:169
128. Zheng J, Post CB (1993) J Mag Res B 101:262
129. Lippens GM, Cerf C, Hallenga K (1992) J Mag Res 99:268
130. Campbell AP, Sykes BD (1993) Annu Rev Biophys Biomol Struct 22:99
131. Lian LY, Barsukov IL, Sutcliffe MJ, Sze KH, Roberts GCK (1994) Methods Enzymol 239:657
132. Moseley HNB, Curto EV, Krishna NR (1995) J Mag Res B 108:243
133. Arepalli SR, Glaudemans CPJ, Daves GD, Kovac P, Bax AJ (1995) J Mag Res B 106:195
134. Vincent SJF, Zwahlen C, Post CB, Burgner JW, Bodenhausen G (1997) Proc Nat Acad Sci USA 94:4383
135. Haselhorst T, Espinosa JF, Jiménez-Barbero J, Sokolowski T, Kosma P, Brade H, Brade L, Peters T (1999) Biochemistry 39:12778
136. Kalk A, Berendsen HJC (1976) J Mag Res 24:343
137. Baleja JD, Moult J, Sykes BD (1990) J Mag Res 87:375
138. Borrelli S, Hossany RB, Findlay S, Pinto BM (2006) Am J Immunol 2:73
139. Beenhouwer DO, Valadon P, May R, Scharff M (2000) Peptide mimicry of the polysaccharide capsule of *Cryptococcus neoformans*. In: Cunningham MW, Fuji-

nami RS (eds) Molecular mimicry, microbes, and autoimmunity, chap 11. ASM Press, Washington, DC
140. Nakouzi A, Valadon P, Nosanchuk J, Green N, Casadevall A (2001) Infect Immun 69:3398
141. Oscarson S, Alpe M, Svahnberg P, Nakouzi A, Casadevall A (2005) Vaccine 23:3961
142. Casadevall A (1995) Infect Immun 63:4211
143. Vecchiarelli A, Casadevall A (1998) Res Immunol 149:321
144. Valadon P, Nussbaum G, Boyd LF, Margulies DH, Scharff MD (1996) J Mol Biol 261:11
145. Valadon P, Nussbaum G, Oh J, Scharff MD (1998) J Immunol 161:1829
146. Young ACM, Valadon P, Casadevall A, Scharff MD, Sacchettini JC (1997) J Mol Biol 274:622
147. Kabat EA, Wu TT (1991) J Immunol 147:1709
148. McLean GR, Torres M, Elguezabal N, Nakouzi A, Casadevall A (2002) J Immunol 169:1379
149. Nussbaum G, Cleare W, Casadevall A, Scharff MD, Valadon P (1997) J Exp Med 185:685
150. Pirofski L (2001) Trends Microbiol 9:445
151. Kaur KJ, Khurana S, Salunke DM (1997) J Biol Chem 272:5539
152. Kaur KJ, Jain D, Goel M, Salunke DM (2001) Vaccine 19:3124
153. Jain D, Kaur K, Sundaravadivel B, Salunke DM (2000) J Biol Chem 275:16098
154. Moothoo DN, Naismith JH (1998) Glycobiology 8:173
155. Jain D, Kaur KJ, Salunke DM (2001) Biophys J 80:2912
156. Jain D, Kaur KJ, Salunke DM (2001) Biochemistry 40:12059
157. Jain D, Kaur KJ, Goel M, Salunke DM (2000) Biochem Biophys Res Commun 272:843
158. Goel M, Krishnan L, Kaur S, Kaur KJ, Salunke DM (2004) J Immunol 173:7358
159. Böhm HJ (1992) J Comput Aided Mol Des 6:593
160. Dam TK, Brewer CF (2002) Chem Rev 102:387
161. Murali R, Kieber-Emmons T (1997) J Mol Recognit 10:269
162. Blaszczyk-Thurin M, Murali R, Westerink MAJ, Steplewski Z, Co MS, Kieber-Emmons T (1996) Protein Eng 9:447
163. Ziebell MR, Prestwich GD (2004) J Comp Aided Mol Des 18:597
164. Dokurno P, Bates PA, Band HA, Stewart LMD, Lally JM, Burchell JM, Taylor-Papadimitriou J, Snary D, Sternberg MJE, Freemont PS (1998) J Mol Biol 284:713
165. Morris GM, Goodsell DS, Halliday RS, Huey R, Hart WE, Belew RK, Olson AJ (1998) J Comput Chem 19:1639
166. Houston DR, Shiomi K, Arai N, Omura S, Peter MG, Turberg A, Synstad B, Eijsink VGH, van Aalten DMF (2002) Proc Nat Acad Sci USA 99:9127
167. van Aalten DMF, Komander D, Synstad B, Gaseidnes S, Peter MG, Eijsink VGH (2001) Proc Nat Acad Sci USA 98:8979
168. Shiomi K, Arai N, Iwai Y, Turberg A, Koelbl H, Omura S (2000) Tetrahedron Lett 41:2141
169. Arai N, Shiomi K, Yamaguchi Y, Masuma R, Iwai Y, Turberg A, Koelbl H, Omura S (2000) Chem Pharm Bull 48:1442
170. Rao FV, Houston DR, Boot RG, Aerts JMFG, Hodkinson M, Adams DJ, Shiomi K, Omura S, van Aalten DMF (2005) Chem Biol 12:65
171. Bompard-Gilles C, Rousseau P, Rougé P, Payan F (1996) Structure 4:1441
172. Qian M, Haser R, Buisson G, Duée E, Payan F (1994) Biochemistry 33:6284
173. Wiegand G, Epp O, Huber R (1995) J Mol Biol 247:99
174. Strobl S, Maskos K, Wiegand G, Huber R, Gomis-Rüth FX, Glockshuber R (1998) Structure 6:911

175. Vallée F, Kadziola A, Bourne Y, Juy M, Rodenburg KW, Svensson B, Haser R (1998) Structure 6:649
176. Desmyter A, Transue TR, Arbabi Ghahroudi M, Dao Thi MH, Poortmans F, Hamers R, Muyldermans S, Wyns L (1996) Nat Struct Biol 3:803
177. Transue TR, De Genst E, Arbabi Ghahroudi M, Wyns L, Muyldermans S (1998) Proteins 32:515
178. Jerne NK (1974) Ann Immunol 125C:373 (Paris)
179. Evans SV, Rose DR, To R, Young M, Bundle DR (1994) J Mol Biol 241:691
180. Young NM, Gidney MAJ, Gudmundsson BME, MacKenzie CR, To R, Watson DC, Bundle DR (1999) Mol Immunol 36:339
181. Mond JJ, Lees A, Snapper CM (1995) Annu Rev Immunol 13:655
182. Gulati S, McQuillen DP, Sharon J, Rice PA (1996) J Infect Dis 174:1238
183. Beninati C, Arseni S, Mancuso G, Magliani W, Conti S, Midiri A, Biondo C, Polonelli L, Teti G (2004) J Immunol 172:2461
184. Poljak RJ (1994) Proc Nat Acad Sci USA 91:1599
185. St. Hilaire PM, Meldal M (2000) Angew Chem Int Ed 39:1162
186. Meldal M, Christiansen-Brams I, Christensen MK, Mouritsen S, Bock K (1994) In: Bock K, Clausen H (eds) Complex carbohydrates in drug research: Structural and functional aspects. Munksgaard, Copenhagen, p 153
187. Christensen MK, Meldal M, Bock K, Cordes H, Mouritsen S, Elsner H (1994) J Chem Soc Perkin Trans 1:1299
188. Lin CC, Kimura T, Wu SH, Weitz-Schmidt G, Wong CH (1996) Bioorg Med Chem Lett 6:2755
189. Xian M, Fatima Z, Zhang W, Fang J, Li H, Pei D, Loo J, Stevenson T, Wang PG (2004) J Comb Chem 6:126
190. Ying L, Liu R, Zhang J, Lam K, Lebrilla CB, Gervay-Hague J (2005) J Comb Chem 7:372
191. Debenham SD, Snyder PW, Toone EJ (2003) J Org Chem 68:5805
192. Halkes KM, St. Hilaire PM, Crocker PR, Meldal M (2003) J Comb Chem 5:18
193. Borrelli S, Johnson MA, Hossany RB, Pinto BM (2007) Peptide mimics of bacterial polysaccharides: potential for discriminating vaccines. In: Roy R (ed) Carbohydrate vaccines. ACS Symp Ser. American Chemical Society, Washington, DC, in press
194. Maitta RW, Datta K, Lees A, Belouski SS, Pirofski L (2004) Infect Immun 72:196
195. Kieber-Emmons T, Monzavi-Karbassi B, Wang B, Luo P, Weiner DB (2000) J Immunol 165:623
196. Qiu JX, Marcus DM (1999) J Neuroimmunol 100:58
197. Cunto-Amesty G, Kieber-Emmons T (2000) Exploiting molecular mimicry in targeting carbohydrate antigens. In: Cunningham MW, Fujinami RS (eds) Molecular mimicry, microbes, and autoimmunity, chap 13. ASM Press, Washington, DC
198. Cunto-Amesty G, Luo P, Monzavi-Karbassi B, Lees A, Kieber-Emmons T (2001) Vaccine 19:2361
199. Monzavi-Karbassi B, Kieber-Emmons T (2001) Biotechniques 30:170

Top Curr Chem (2008) 273: 117–138
DOI 10.1007/128_2007_145

Published online: 13 October 2007

Studies on the Conformational Features of Neomycin-B and its Molecular Recognition by RNA and Bacterial Defense Proteins

Juan Luis Asensio[1] (✉) · Agatha Bastida[1] · Jesús Jiménez-Barbero[2] (✉)

[1]Instituto de Química Orgánica, CSIC, Juan de la Cierva 3, 28006 Madrid, Spain
iqoa110@iqog.csic.es

[2]Centro de Investigaciones Biológicas, CSIC, Ramiro de Maeztu 9, 28040 Madrid, Spain
jjbarbero@cib.csic.es

Abstract According to NMR and molecular dynamics simulations, the conformational behavior of natural aminoglycosides is characterized by a remarkable flexibility, with different conformations, even *non-exo-anomeric* ones, in fast exchange. Very probably, this feature allows the adaptation of these ligands to the spatial and electronic requirements of different receptors. The large diversity of structures adopted by aminoglycosides in the binding pocket of the different RNA receptors and the distinct enzymes involved in bacterial resistance are consistent with this view. This conformational diversity can, in certain favorable cases, be exploited in the design of new antibiotic derivatives not susceptible to enzymatic inactivation, by designing tailor-made conformationally locked aminoglycosides.

1 Introduction

Aminoglycosides of the neomycin family are a class of clinically relevant antibiotics. These drugs bind to the decoding region aminoacyl-tRNA site (A-site), inducing codon misreading and inhibiting translocation, eventually resulting in cell death [1–4]. From a chemical point of view, aminoglycosides are polycationic oligosaccharides carrying several hydroxyl and amino functional groups. Most include one or more amino sugars attached to a 2-deoxystreptamine ring (ring II in Fig. 1).

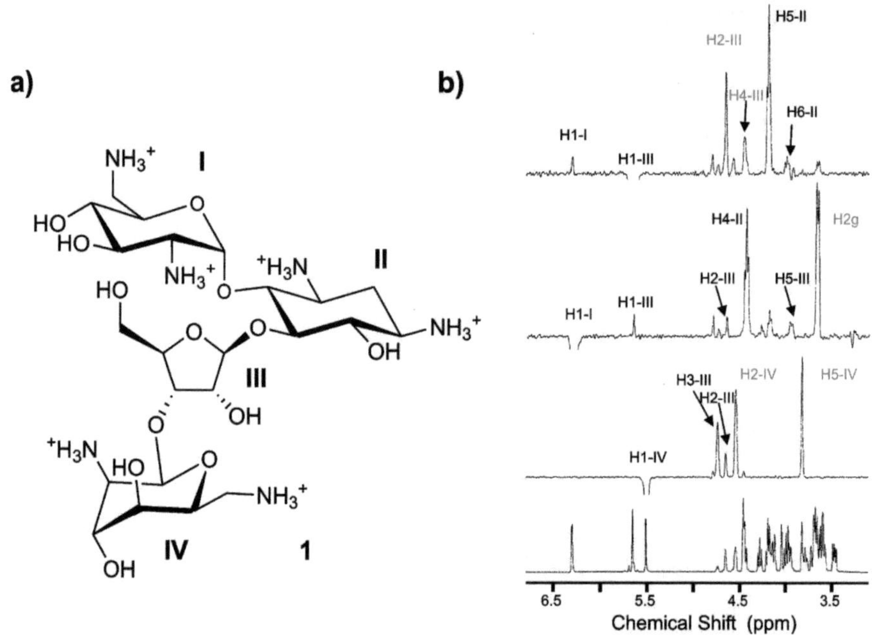

Fig. 1 **a** Schematic representation of Neomycin-B along with the numbering employed for the different sugar units. **b** Selective 1-D NOE experiments with the 1D-DPFGSE NOE pulse sequence, corresponding to the inversion of $H1_{III}$, $H1_{I}$ and $H1_{IV}$ (from *top* to *bottom*)

Several years after the introduction of aminoglycosides in human antibacterial chemotherapy, resistant organisms began to appear. Indeed, the emergence of bacterial resistance to the major classes of antibiotics has become a serious problem over recent years [5, 6]. Acquired resistance to aminoglycoside antibiotics can occur via three different mechanisms: mutation of the ribosomal target, reduced permeability for antibiotics and enzymatic modification of the drugs, thus leading to inactivation. The most prevalent source of clinically relevant resistance is conferred by the third mechanism, enzymatic inactivation of the drugs [5, 6].

The three-dimensional structure of the aminoglycosides plays an essential role in their interaction with both RNA and the enzymes involved in the antibiotic inactivation and thus determines their biological function. A proper understanding of the different factors that govern aminoglycoside-RNA/protein interactions requires a detailed knowledge of the three-dimensional structure, and conformational properties of these oligosaccharides in both the free and bound states. In this context, the rigidity versus flexibility of natural glycosides is an issue of great interest, because it concerns the potential adaptability of these ligands to the spatial and electronic requirements of the different receptors [7–11]. Although absolute rigidity can obviously be ruled out, the views concerning the degree of flexibility present in natural oligosaccharides have evolved from the acceptance of only very restricted fluctuations around a preferred conformation to wide variability around the glycosidic linkages with different conformations in exchange depending on their nature. Thus, the presence of anti-Φ and anti-Ψ conformations (Φ/Ψ (H–C–O–C) ca. 180°) in solution for some particular linkages was first suggested by molecular mechanics calculations, and later detected by NMR spectroscopy [12, 13]. Nevertheless, it has to be pointed out that in all cases, these anti Φ-conformations were populated not higher than 5%.

Herein, we describe conformational features of neomycin-B in the free state [14, 15]. The structures adopted by neomycin-B and related aminoglycosides bound to RNA and enzymes involved in bacterial resistance, according to X-ray data, will also be discussed. Interestingly, the detailed comparison between the antibiotic free and RNA/protein bound conformations suggests a possible strategy for the design of new aminoglycosides with improved activity against resistant bacteria. This strategy will be presented in the final section of this manuscript.

2
Structural Analysis of Neomycin-B in Solution

In order to determine the influence of protonation on the conformational preferences of neomycin-B, the structural analysis of the antibiotic was carried out at two different pH values, 4.7 and 9.7, corresponding to the fully protonated and neutral states respectively.

2.1
pH 4.7 (the Fully Protonated State)

As a first step, selective 1D-NOEs experiments were carried out at 313 K and pH 4.7. The branched nature of neomycin-B (Fig. 1) allowed the measurement of an unusually large number of structurally relevant NOEs. In addition, ^3J values were measured for the ring III. The analysis of the couplings for unit IV unambiguously shows that the $^1C_4$(L) conformer (with three axial and two equatorial substituents) is the most important one in water(> 98%). Thus, the 1H-NMR signals for the H1–H4 protons are broad singlets, in agreement with a gauche-type arrangement of the vicinal proton pairs. In addition, a very strong H1/H5 intraresidue NOE indicates a syn-diaxial relationship for this proton pair.

Careful inspection of all the NOE and J data immediately suggests the existence of a significant degree of internal mobility in the tetrasaccharide; no single structure can satisfy all the constraints simultaneously. In order to obtain an experimentally derived ensemble, 80 ns ("in vacuo", $\varepsilon = 80$) MD-tar[11] simulations (molecular dynamics with time-averaged restraints) were carried out by including five coupling constants and 14 experimental distances as time-average restraints with the AMBER 5.0 program [17].

The MD-tar distribution of conformers obtained for every particular glycosidic linkage is shown in Fig. 2a, superimposed on the MM3* steric maps previously calculated. It can be observed that for the I/II linkage, a major population is centered around $\phi/\psi = -55°/-45°$. A very minor population around $\phi/\Psi = -25°/30°$ was also detected. In contrast, the IV/III linkage is characterized by a high degree of flexibility, with two different conformations almost equally populated located at $\phi/\Psi = -50°/-50°$ and $\phi/\Psi = -50°/45°$. Nevertheless, both glycosidic linkages exhibit common features: ϕ values are

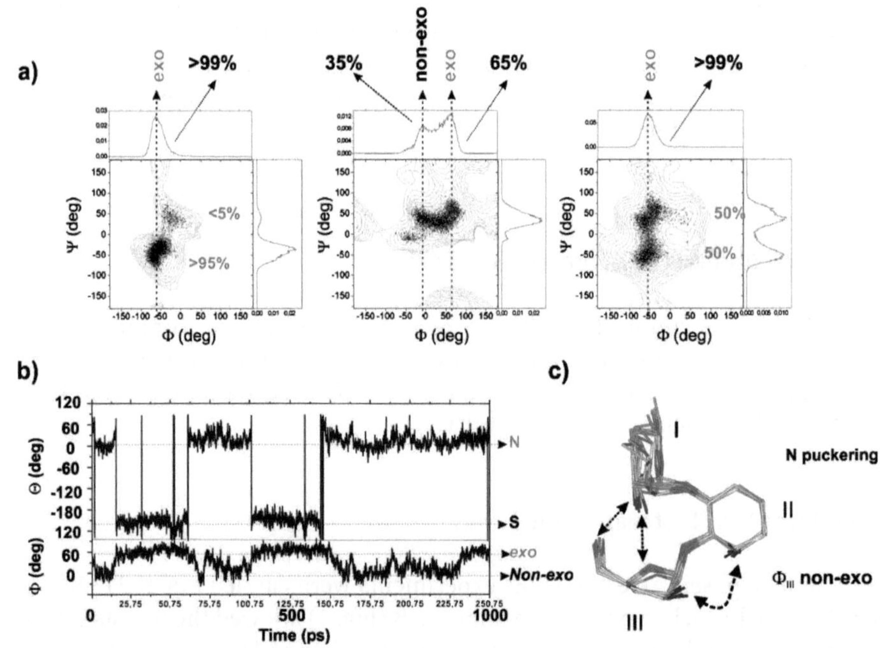

Fig. 2 **a** Experimentally derived MD-tar distributions obtained from a 80 ns length simulation for the I/II (*left*), III/II (*middle*) and IV/III (*right*) linkages at pH 4.7. **b** Variations in ϕ_{III} (*bottom*) and the puckering Phase angle of unit III (*top*) during a few picoseconds of the MD-tar simulations. Deviations from the *exo-anomeric* region are strongly correlated with the ring puckering. **c** Snapshot, taken from the solvated MD simulation, corresponding to neomycin-B with unit III in the "N" state. Polar contacts between the non-vicinal units I and III are shown in the *upper part*. In addition, the $H2_{III}$–$H6_{II}$ NOE, characteristic of non-exo-anomeric ϕ_{III} angles is shown

scattered around the *exo-anomeric* region and conformational flexibility is mainly restricted to the aglyconic ψ torsion. This constitutes the usual behavior in all natural O-glycosides described so far. A totally different behavior is observed for the III/II linkage. In this case, two populations characterized by different ϕ values (60°/30° and −10°/30°) were detected. In contrast to common behavior, this glycosidic linkage shows a larger degree of mobility around ϕ than around ψ. Moreover, MD-tar simulations indicate that a large percentage of the population (> 30%) is located in *non-exo-anomeric* regions with ϕ ca. −10°. Indeed, the occurrence of a *non-exo-anomeric* orientation around ϕ_{III} is experimentally granted by a medium NOE $H2_{III}/H6_{II}$, exclusive of this conformational region. That is, the existence of a short $H2_{III}/H6_{II}$ average distance can never be explained without assuming a significant deviation from the *exo-anomeric* region for ϕ_{III} (the $H2_{III}/H6_{II}$ distance for the *exo-anomeric* region is longer than 4.1 Å).

Interestingly, there is a strong correlation between the ring III conformation and the glycosidic torsion ϕ_{III} according to the MD-tar data. The experimentally derived puckering distributions show the presence of two different conformational regions populated in solution with phase angles around 10° (60–70%) and 160° (30–40%) (herein referred as "N" and "S" regions respectively). Thus, when unit III is in the "S" state, ϕ_{III} adopts values for the *exo-anomeric* region (Fig. 2b). In contrast, for the "N" region, ϕ_{III} presents significant deviations from the *exo-anomeric* values. Detailed inspection of the structures shows the presence of a bifurcated hydrogen bond involving the amino group at position 2 of unit I as a donor and both O4 and O5 in unit III as acceptors (Fig. 2c). This polar contact, between non-vicinal residues, is responsible for the observed shift in ϕ_{III} towards *non-exo-anomeric* regions. In fact, according to "in vacuo" MD runs with $\varepsilon = 80$, all glycosidic torsions are in agreement with the *exo-anomeric* affect, independently of the ring III conformation. In addition, trajectories collected in presence of explicit water molecules and counterions indicated that the polar contacts between the non-vicinal units I and III are selectively disrupted by solvent only for the "S" puckering, but not for the "N". Thus, the different effect of the solvent on both species is at the origin of the observed correlation between the ϕ_{III} angle and the five-member ring puckering. As a final test, solvated MD-tar simulations of the non-branched 2-6 dideoxy-2,6 diamino-L-Ido β(1–3) Ribβ(1–5) 2-deoxy-Strp fragment of neomycin-B (rings II, III and IV) employing only theoretical J restrains for the ribose were carried out. The obtained results for this linear trisaccharide conclusively show that in the absence of branching the overall rigidity around ϕ_{III} increases, independently of the puckering distribution considered for unit III, and now all torsion values are consistent with the *exo-anomeric* effect.

The conformational properties of the IV/III linkage also seem to be correlated with the structure of unit III (Fig. 3a). Thus, ψ angle adopts mainly positive values (around +45°) for a "S" ring III and negative (around −50°)

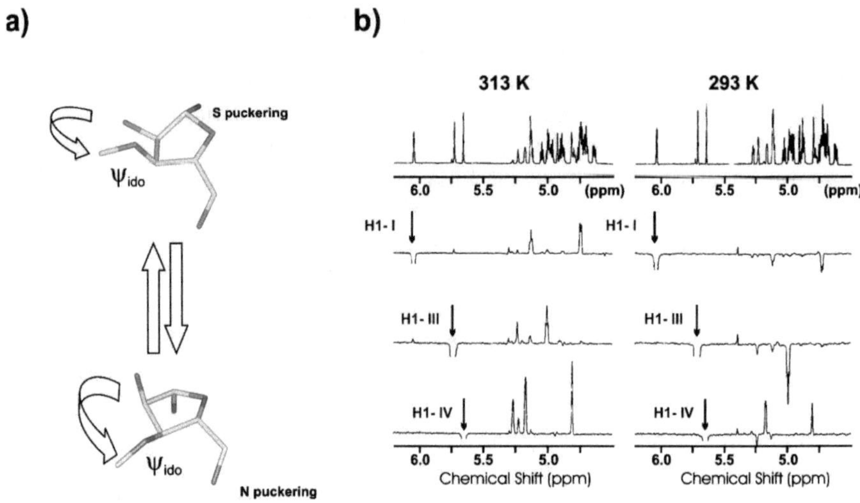

Fig. 3 **a** 1 : 3 syn-diaxial interactions established by $OH2_{III}$ and $OH5_{III}$ with the vicinal $OH3_{III}$ group are dependent on the ring III conformation. **b** Dependency exhibited by the intraresidue NOEs on temperature for the different sugar units. Selective NOEs obtained upon inversion of the three anomeric protons at 313 K (*left*) and 293 K (*right*)

for a "N" conformation (although it is significantly more flexible for the later case). It is well known that conformational preferences around ψ angle in natural O-glycosides are modulated by 1 : 3 syn-diaxial interactions [18] with vicinal OH groups (Fig. 3a). According to this view, conformational populations characterized by negative ψ values (Fig. 3a) would be destabilized by a 1 : 3-type contact with the pseudoequatorial $OH2_{III}$ only for the "S" puckering. This contact does not exist for a "N" conformation where $OH2_{III}$ is pseudoaxial and therefore a larger flexibility is expected, in agreement with the MD-tar results.

According to the MD-tar simulations the IV/III linkage has the larger degree of internal mobility in the tetrasaccharide, followed by the III/II bond. This point is further supported by the dependency exhibited by the intraresidue NOEs on temperature for the different sugar units. Figure 3b shows selective NOEs obtained on inversion of the three anomeric protons at 313 K (left) and 293 K (right). It can be observed that, at 313 K, the three sugar units (I, III and IV) present positive NOEs (high mobility). In contrast, at 293 K, an inversion in the sign of the intraresidue contacts is observed for the I and III units, but not for the ring IV. This observation is consistent with a shorter local correlation time for this terminal residue. Unit III also exhibits a large degree of flexibility according to the MD-tar data. However, in this case, internal mobility around the glycosidic linkage is couple to the "N"/"S" exchange in the furanose ring and influenced by remote contacts with unit I. Therefore, a larger energy barrier and a slower conformational exchange would be expected.

The conformational behavior of neomycin-B was dependent on pH. At pH 9.7 (neutral state), the NOE data were consistent with an increase in the population around the anti Ψ regions for the I/II and III/II linkages (Fig. 4). However, the deviations of Φ_{III} from the exoanomeric region were less pronounced that those observed at pH 4.7, indicating that the polar interactions between units III and I are weaker for the neutral antibiotic. A slightly different conformational behavior for the IV/III linkage was also suggested. Differences between both pH values are evident for the three glycosidic linkages (Fig. 4).

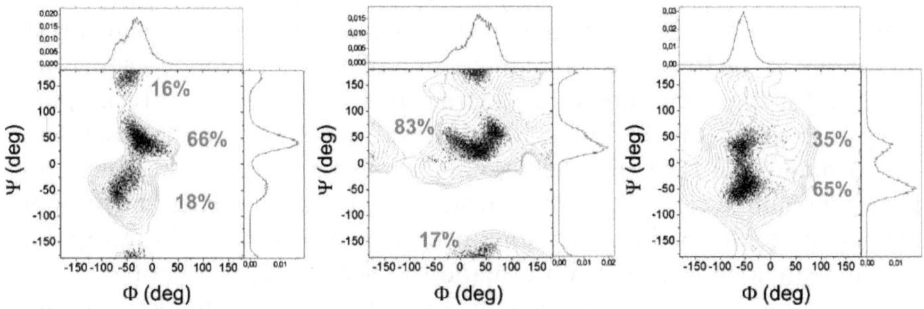

Fig. 4 Experimentally derived MD-tar distributions obtained from a 80 ns length simulation for the I/II (*left*), III/II (*middle*) and IV/III (*right*) linkages at pH 9.7

In conclusion the conformational analysis of neomycin-B in the free state unequivocally showed that flexibility in aminoglycosides is not restricted to the aglyconic Ψ angle. In certain cases, Φ can also undergo conformational fluctuations, even adopting non-exo-anomeric orientations. To the best of our knowledge, these antibiotics represent the first reported case of occurrence of a large population of non-exo-anomeric conformers in solution. Many relevant saccharides, such as the blood group antigens, include sugar units glycosidated in contiguous positions. It has been assumed that this pattern of substitution increases the overall rigidity of the ligand due to steric contacts between non-vicinal substituents [19, 20]. Our results indicate that this cannot be considered as a general rule, and the precise conformational behaviour of the oligosaccharide depends on the configuration and chemical nature of its different sugar units. For neomycin-B, polar contacts between non-vicinal sugar moieties (I and III) lead to an enhanced flexibility. The large internal mobility exhibited by natural neomycin-B in the free state probably has a significant influence on its molecular recognition by RNA and proteins. Thus, it probably makes possible the adaptation of the ligand to RNA/protein binding pockets of different topology with a small ΔH penalty. In addition, in those cases where the aminoglycoside flexibility is severely restricted upon binding to a given receptor, an unfavourable ΔS contribution to the globals ΔG is expected.

2.2
Structure and Flexibility of Aminoglycoside Antibiotics Bound to the A-site RNA

The 3D structures of several aminoglycoside/A-site RNA complexes have been described recently by both X-ray and NMR methods [21–26]. Thus, according to these data the structure of the neomycin-B related antibiotic paromomycin in the RNA-binding pocket can be described in terms of a single *exo-anomeric* conformation for the three glycosidic linkages. In addition, the three glycosidic linkages adopt a sin-Ψ (Ψ in the $-50°/+50°$ range) arrangement. Interestingly, in the aminoglycoside RNA-bound state, ring III exhibits a puckering in the "South" region of the pseudorotational path, which is the minor one in the free state. Close inspection of molecular models shows that the global shape of the antibiotic is extremely dependent on the precise puckering of this ring (Fig. 5). In fact, the specific drug-RNA contacts observed in the paromomycin-RNA complex [21–23], involving units III and IV and presumably required for tight binding, would only be feasible when the furanose adopts this conformation. Therefore, neomycin-B/paromomycin recognition by the A-site RNA implies a conformational selection phenomenon.

In addition, to facilitate the adaptation of the ligand to the RNA binding site, the ligand flexibility might have an additional effect on the recognition

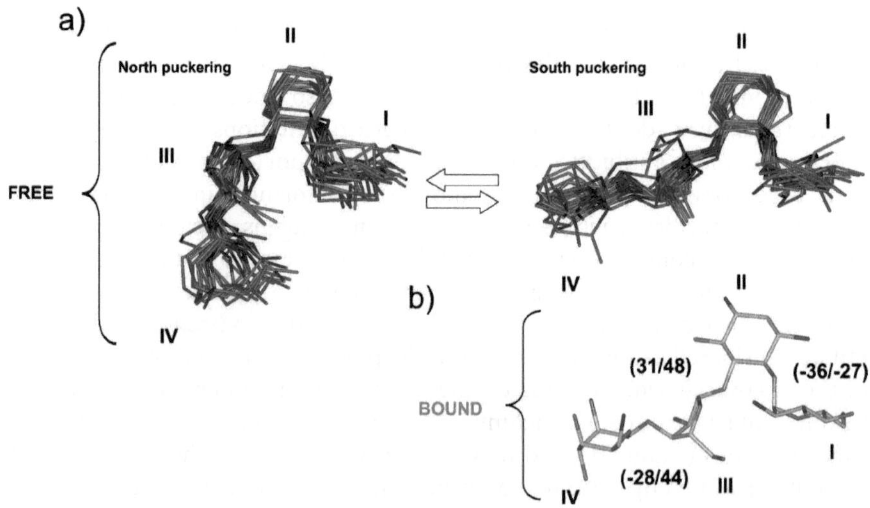

Fig. 5 **a** Neomycin-B conformational families characterized by "N" (*left*) and "S" (*right*) rinbose (ring III) puckering obtained from the MD-tar calculations.The global shape of neomycin-B is extremely dependent on the precise puckering of this unit. Only the scheleton carbon and oxygen atoms are shown for clarity. **b** Structure of the neomycin-B related antibiotic paromomycin bound to the A-site. The Φ/Ψ values for the three glycosidic linkages are indicated

process. Thus, in those cases in which the oligosaccharide internal mobility is severely restricted on binding to RNA, a significant unfavorable entropic contribution to the global ΔG of the process would be expected. The degree of conformational restriction imposed on the ligand by a given RNA receptor can, in principle, be estimated by MD simulations. We have performed this analysis for the complex between ribosomal A-site RNA and neomycin-B. The initial coordinates were built from those reported for the paromomycin/A-site complex derived by both X-ray [21] (pdb accession code 1FJG) and NMR [24–26] methods (pdb accession code 1PBR). After a 5 ns run, deviations from the X-ray experimental coordinates were below 1.2 Å. In contrast, this value is 3.2 Å for the trajectory that employed the NMR coordinates as starting point. Probably, the poorer definition of the RNA phosphate

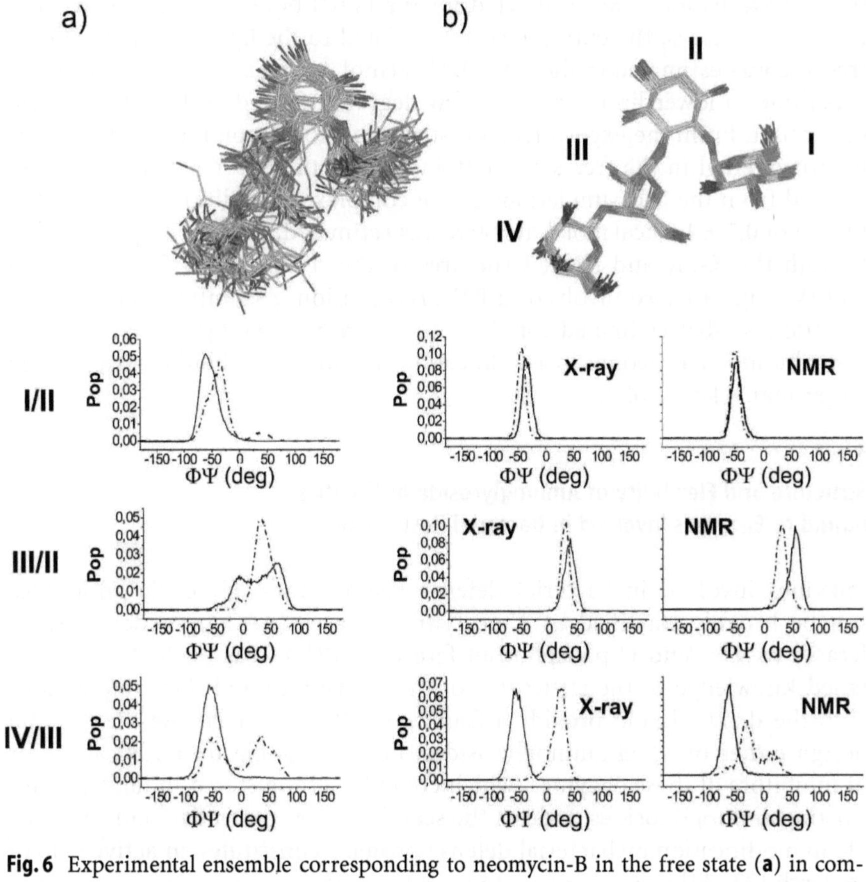

Fig. 6 Experimental ensemble corresponding to neomycin-B in the free state (**a**) in comparison with a theoretical ensemble corresponding to the antibiotic in complex with RNA (**b**) obtained from the MD simulations (starting from both X-Ray and NMR coordinates). Distribution of ϕ (*solid*) and ψ (*dotted*) rotamers in both states are shown for the three glycosidic linkages (from *top* to *bottom*, I/II, III/II and IV/II)

backbone in the initial NMR complex is responsible for this deviation. All the stabilizing sugar/RNA contacts observed in the X-ray structure, and most of those reported for the NMR complex were preserved during the whole trajectories, which in fact predicted a similar degree of conformational restriction in the oligosaccharide on binding to its target RNA fragment.

Figure 6 shows the experimental ensemble corresponding to neomycin-B in the free state (obtained from the MD-tar simulations described above) in comparison with the MD ensemble corresponding to the antibiotic in complex with RNA. These distributions were employed to evaluate the contribution ($T\Delta S$) that rigidification of every glycosidic linkage has to the global ΔG at 300 K. A significant loss of conformational freedom upon binding was predicted, mainly for the III/II and I/II linkages with $T\Delta S$ values of around 1 kcal mol^{-1} per linkage at 300 K. A slightly lower contribution (0.5 ± 0.8 kcal mol^{-1}) was predicted for the IV/III bond. Taking into account the three linkages, the entropic cost associated to the ligand conformational freezing was estimated in the 2.4 ± 3.0 kcal mol^{-1} range. Obviously, this value constitutes a lower limit, as only a few degrees of freedom have been taken in account. From the experimental distribution corresponding to the ring III hydroxymethyl in the free state (obtained from the MD-tar data) and those derived from the MD simulations of the complex, an additional $T\Delta S$ contribution of 0.7 ± 1.0 kcal mol^{-1} at 300 K was estimated. Interestingly, according to both the X-ray and NMR structures of the complex, the OH6 of both I and IV rings are also involved in RNA recognition. Assuming a similar contribution to that estimated for the unit III hydroxymethyl group, the total contribution that neomycin rigidification has to ΔG of binding is probably larger than 4 kcal mol^{-1}.

2.3 Structure and Flexibility of Aminoglycoside Antibiotics Bound to Enzymes Involved in Bacterial Resistance

Enzymes involved in bacterial defense against aminoglycoside antibiotics can be broadly classified as *N*-acetyltransferases (AACs), *O*-adenyl transferases (ANTs) and *O*-phosphotransferases (APHs). It is evident that a detailed knowledge of the structures of these enzymes and their interactions with the drugs should provide a framework that will facilitate the rational design either of novel aminoglycosides, not susceptible to modification, or of inhibitors of these enzymes, which could be administered in tandem with existing aminoglycosides. Indeed, the search for new derivatives not susceptible to modification by bacterial defense proteins constitutes an active field of research [27–34].

The 3D structures of several aminoglycoside-enzyme complexes have been described recently by X-ray methods [35, 36]. Interestingly, in some cases, the 3D structure of the antibiotic in the enzyme binding site significantly differs from that observed in the corresponding aminoglycoside/A-site RNA complex (see Fig. 7). For example, *Staphylococcus aureus* ANT(4′) selects a high energy conformation of the ligand kanamycin [37]. In particular, the Glcα(1–4)-2-deoxy streptamine fragment (units I and II), present in all aminoglycosides with different patterns of amination/hydroxylation at the glucose unit (I), adopts an anti-Ψ geometry (Φ/Ψ, – 21/151). In contrast, for tobramycin and paromomycin, the analogous glycosidic linkage is defined by a syn-Ψ conformation ($\Phi/\Psi \sim$ – 40/– 30) in the RNA-bound state. The "syn-Ψ" geometry also constitutes the most populated minimum for the free antibiotics, according to NMR data. The drug conformation selected by ANT(4′) represents a high-energy minimum that must be stabilized by specific aminoglycoside-protein interactions. In a similar way, the X-ray structure of *Mycobacterium tuberculosis* AAC(2′) in complex with ribostamycin shows that the Ribβ(1–5)-2-deoxy streptamine fragment (units III and II) also adopts a high energy "anti-Ψ" conformation (Φ/Ψ = – 8/– 173) [38]. This geometry is significantly different to the *bioactive* "syn-ψ" orientation adopted by the same glycosidic bond in the paromomycin/A-site complex [21–26]. An additional key example of conformational differences of aminoglycosides between the RNA- and protein-bound states is provided by the recently reported structure of the *Salmonella enteritidis* AAC(6′) in complex with ribostamycin [39]. In this case, the Ribβ(1–5)-2-deoxy streptamine fragment (units III and II) of the oligosaccharide presents a rather unusual non-exo-anomeric conformation (characterized by a Φ angle of –5°). In addition, the furanose ring (unit III) is bound in a C2-endo conformation, in

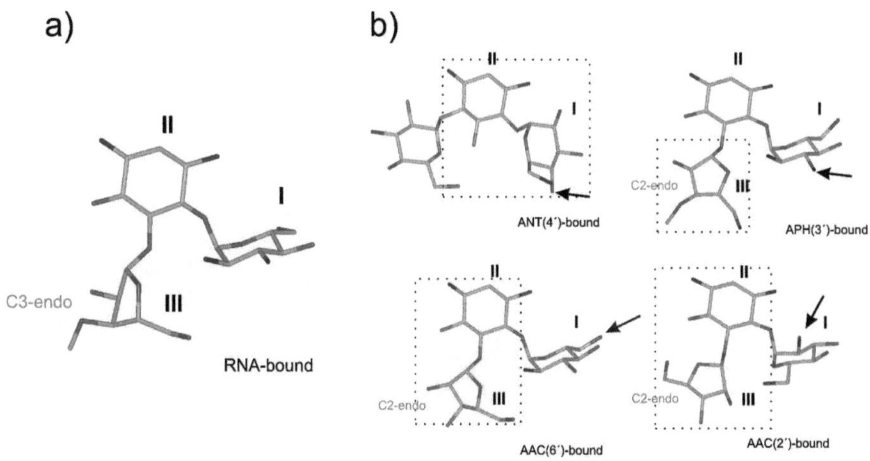

Fig. 7 Schematic representation of the conformational differences (*highlighted with a dotted box*) exhibited by aminoglycosides in the binding pockets of RNA (**a**) and some of the enzymes involved in bacterial resistance (**b**). The glucose, 2-deoxy-streptamine and ribose units are numbered as I, II and III respectively

contrast with the C3-endo geometry selected by the ribosomal RNA. The fact that a significant deviation from the favored exo-anomeric conformation is observed for a ribose unit with a "N" puckering is in agrement with the conformational properties of free neomycin-B previously described. A similar furanose puckering is also recognized by the *Enterococcus faecalis* APH(3′) enzyme according to the X-ray structure of its complex with neomycin-B [40].

3 Exploring Conformational Restriction as a new Strategy to Overcome Bacterial Resistance

The different 3D-shapes adopted by aminoglycosides in the RNA- and enzyme-bound states suggest a possible structure-based chemical strategy to obtain antibiotics with better activity against resistant bacteria. Assuming that, in these cases, some degree of conformational distortion of the substrates is required for enzymatic activity, it should be possible to design a conformationally locked oligosaccharide that still retains antibiotic activity, but that is not susceptible to enzymatic inactivation (Fig. 8) [41].

In order to test this hypothesis, we have designed and synthesized several neomycin-B analogs locked in the RNA-bound (*bioactive*) structure of

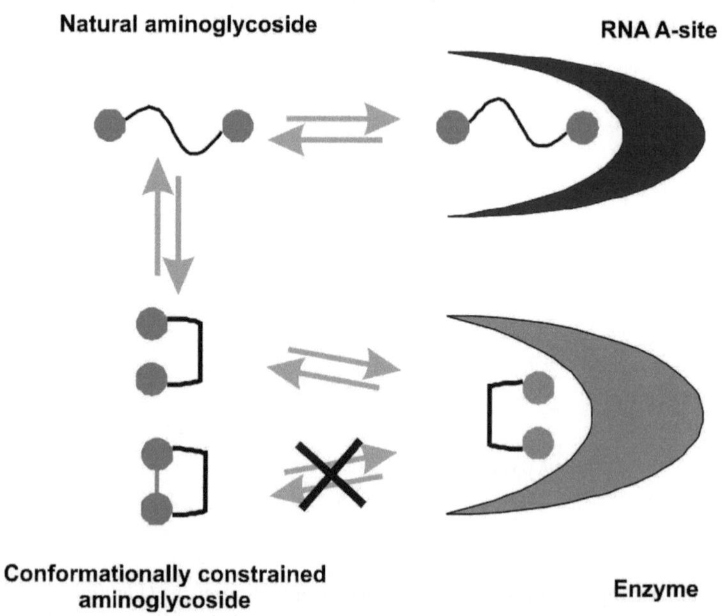

Fig. 8 Schematic representation of the proposed structure-based chemical strategy to overcome bacterial resistance

the antibiotic. Thus, we have studied the effect of structural preorganization on the strength of the aminoglycoside–RNA interaction. In a second step, we have determined the susceptibility of the conformationally locked derivatives to enzymatic inactivation, employing the enzymes *Staphylococcus aureus* ANT(4′), *Mycobacterium tuberculosis* AAC(2′) and *Enterococcus faecalis* APH(3′) as model systems.

The design of our conformationally constrained aminoglycosides is based on the structural information available for the RNA-paromomycin complex [21–26]. It is known that positions $O5_{III}$ and $N2_{I}$ are hydrogen bonded within the ribosome bound state. Thus, in our analogs, this interaction has been replaced by either a direct covalent bond, as in **2**, a two-methylene bridge, as in **3**, or by a salt bridge, as in **4** (see Fig. 9).

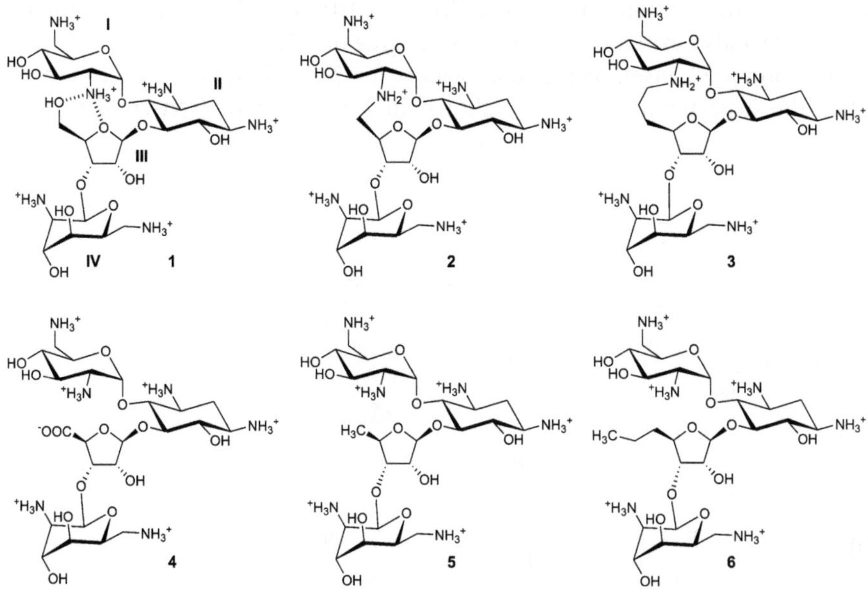

Fig. 9 Schematic representation of the target conformationally locked aminoglycosides (**2–4**) and the deoxy derivatives (**5**, **6**). The numbering employed for the different sugar units is indicated for neomycin-B (compound 1)

It should be pointed out that $OH5_{III}$ is involved in RNA recognition [21–26] and might also have an influence on the enzymatic inactivation of the antibiotics. Since cyclization of neomycin-B to give **2** or **3** required the removal of this OH group, the deoxy- derivatives **5** and **6** were also synthesized as controls, in order to isolate the effect of rigidification.

The structural features of derivatives **2–4** were analyzed employing a methodology similar to that previously described for neomycin-B (**1**). Thus, selective 1D-NOE experiments were carried out at 313 K and pH 5.0. In addition, ^3J values were measured for all the sugar units. In order to get an experimentally derived ensemble, MD-tar simulations [16] were carried out by including all the experimental NOE-derived distances and J data as time-average restraints with the AMBER 5.0 [17] program, as described [14, 15]. This analysis showed that cyclization of neomycin-B to give **2–4** did not produce significant distortions in any of the three pyranose rings. According to the NMR and MD-tar data, **2** was extremely rigid, with minor fluctuations around the glycosidic linkages und a unique puckering for the ribose ring. Analog **3** showed a larger degree of flexibility with a major (> 80%) "gt" ($\omega_{O\text{–}C4\text{–}C5\text{–}C6} = 60°$) and a minor (< 20%) "gg" ($\omega_{O\text{–}C4\text{–}C5\text{–}C6} = -60°$) orientations for the ribose lateral chain and two different puckerings for the furanose ring, characterized by phase angles of 25° ("North") and 160° ("South") and relative populations of 40 : 60 respectively. Indeed, **3** encompasses relatively large fluctuations around the I/II and III/II glycosidic linkages. Finally, derivative **4** also exhibited a significant degree of flexibility.

The obtained ensembles of conformers for **2–4** that quantitatively reproduce the experimental data are shown in Fig. 10a. These ensembles were

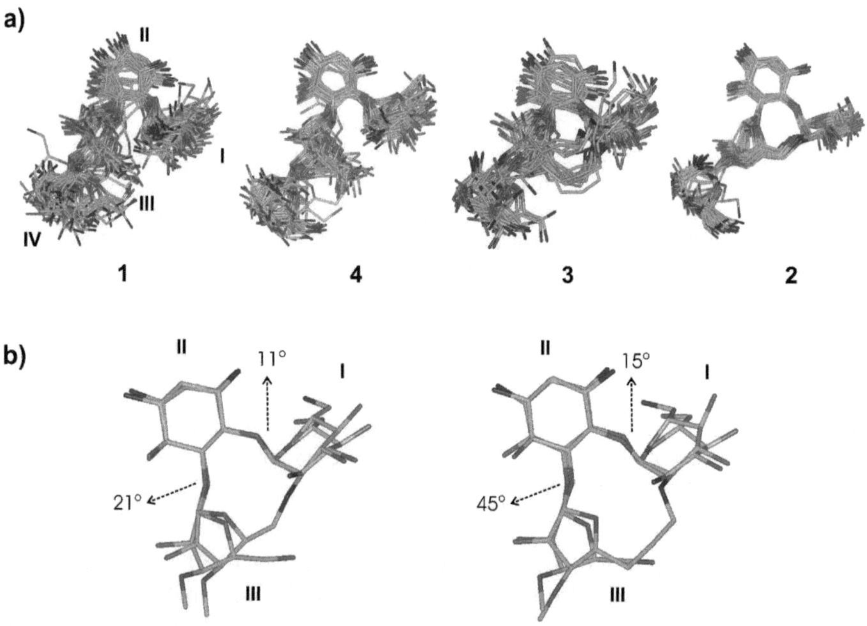

Fig. 10 **a** MD-tar ensembles of conformers for **2–4**. The experimental ensemble corresponding to **1** in the free state (*reft*) is shown for comparison. **b** Structures representative of the main conformational families present in solution for neomycin-B derivatives **2** (*left*) and **3** (*right*) superimposed on the X-Ray structure of paromomycin in the complex with ribosomal RNA, according to X-Ray data. The maximum Φ/Ψ deviations (for each glycosidic linkage) is shown. Unit IV is omitted for simplicity

compared with that obtained for the natural antibiotic **1** in the free (experimentally derived from the NMR data) [14, 15]. As expected, analogs **2–4** are more rigid than **1** in the free state. The RMSD for heavy atom superimposition for the ensemble corresponding to **1** in the free state is 2.1 Å. The corresponding values for **3** and **4** are 1.6 Å. The conformational restriction imposed by covalently bonding $C5_{III}$ to $N2_{I}$ in **2** is especially severe, with an RMSD value of 0.9 Å. Interestingly, the locked aminoglycosides **2–4** closely resemble the bioactive conformation of the natural antibiotic. The most representative minimum energy conformations present in solution for **2** and **3**, superimposed on the X-ray structure of paromomycin in complex with A-site RNA, are shown in Fig. 10b. For **2**, deviations in Φ/Ψ values with respect to the paromomycin bound structure are less than 11° and 21° for the I/II and III/II linkages respectively. The corresponding deviations for **3** are larger (15° and 45°). However, the global shape of the molecule is very similar to that observed for paromomycin in the RNA bound state, as shown by the low RMSD values for heavy atom superimposition (0.5 and 0.7 Å for **2** and **3** respectively).

As a next step, the binding properties of compounds **1–6** to ribosomal RNA were analyzed employing the fluorescence-based approach, developed independently by Pilch [42] and Hermann [43]. This methodology makes use of a modified 27-mer RNA fragment including the A-site sequence and the fluorescent base analogue 2-aminopurine (2AP) replacing A1492. The natural aminoglycoside neomycin-B binds to the A-site with a Kd value of 0.07 μM in agreement with previously reported values obtained by other methods under comparable experimental conditions [44–46]. The observed Kd values for the macrocyclic derivatives **2** and **3** were 1.11 and 0.9 μM respectively. This represents a 16–13 fold decrease in affinity with respect to natural neomycin-B.

Recently, Tor and co-workers have analysed the interaction between the cyclic analog **2** and the corresponding paromomycin derivative with the A-site sequence and also with the HIV TAR RNA fragment [47]. At pH 7.5, the decrease in affinity for the locked derivatives with respect to the parent aminoglycosides is 22–14 fold for the A-site and 11–2 fold for the HIV TAR sequence. The differences with respect to the natural ligands are significantly reduced at acid pH.

Interestingly, according to our data, the deoxy-derivatives **5** and **6** present affinities for the A-site (Kd values of 0.83 and 0.76 μM respectively) in the same range as that previously measured for **2** and **3** under identical conditions, which strongly suggests that the ΔG contribution resulting from the structural constrain is almost negligible. This is surprising, given the significant lost of conformational entropy associated to A-site binding previously deduced for natural neomycin-B from MD simulations. The entropic benefit derived from the conformational restriction is probably compensated by the loss of enthalpy favorable interactions. In fact, the 3D structure of derivative **2** in complex with the A-site, recently reported by Hermann and col. [48] shows that rings III and IV present a slightly different orientation in the RNA major groove from that observed for natural neomycin-B. As a result the number of ligand-RNA contacts established by these two units is reduced. Therefore, both the structural and binding data are consistent with an enthalpy-entropy compensation phenomenon.

Finally, for compound **4**, the maximum increase in fluorescence observed on complex formation was less than half the value measured for neomycin-B, suggesting a smaller structural change in the receptor upon complex formation. It is known that, in the paromomycin-A-site complex, $OH5_{III}$ is located in a strongly electronegative region of the RNA binding pocket [21–26]. Probably, the substitution of this OH group by a negatively charged carboxylate precludes a deep penetration of the ligand in the major groove. However, despite this modification, the observed affinity for **4** is in the micromolar range (Kd= 1.25 μM).

As a further step, the biological activity of the locked neomycin-B derivatives **2**, **3** and **4** was tested against *E. coli* (DH5α). For **2**, a more extensive testing was performed employing both Gram+ and Gram– bacteria. Although a decrease in activity with respect to the natural compound **1** (3–20 fold depending on the bacteria and compound) was observed, **2** and **3** were still significantly active (with MIC values in the 5–20 μg/ml range). In contrast, **4** showed a poor biological activity. Thus, the obtained results for **2** and **3** showed a good correlation with the binding affinities previously measured for ribosomal RNA and suggested that these locked derivatives might potentially be employed as antibiotics.

In order to analyze the effect that conformational restriction has on the antibiotic enzymatic inactivation, three different enzymes were chosen as model systems: *Staphylococcus aureus* ANT(4′), *Mycobacterium tuberculosis* AAC(2′) and *Enterococcus faecalis* APH(3′). These proteins are representative of the three main families of enzymes that modify aminoglycosides: adenyltransferases, acyltransferases and phosphotransferases. In addition, there is high resolution X-ray structural information available for the three enzymes in complex with several antibiotics.

Staphylococcus aureus ANT(4′) catalyzes the transfer of one adenyl group from ATP to position O-4 of ring I (see Fig. 7b). This transfer leads to a sharp decrease in the drug affinity for its target RNA. According to X-ray studies the oligosaccharide conformation recognized by the RNA and by the enzyme is significantly different, as previously mentioned [37]. If it is assumed that some degree of conformational distortion at the I/II fragment is in fact required for enzymatic activity, the conformational restriction introduced in derivatives **2** and **3** might provide a certain degree of protection against the enzymatic inactivation of the antibiotic.

In order to test this hypothesis, enzymatic reactions were monitored in NMR tubes. ATP (100 μM–3 mM) and the aminoglycosides (100 μM–1.5 mM) were dissolved in phosphate buffer (10 mM phosphate, 2.5–5.0 mM $MgCl_2$,

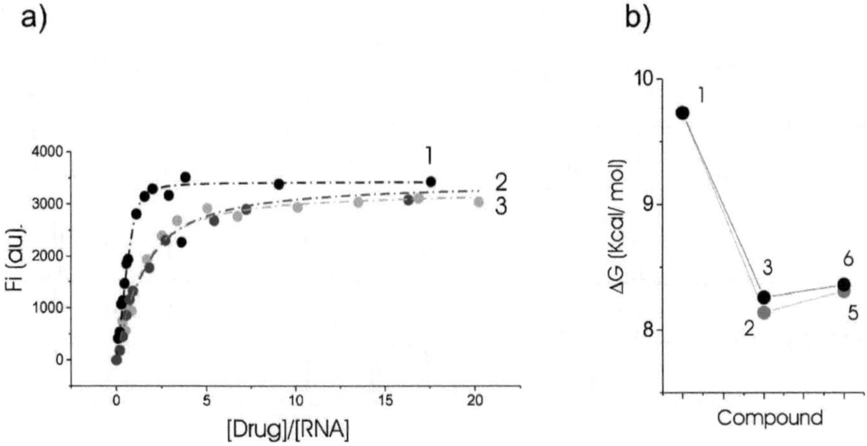

Fig. 11 **a** Some representative binding curves obtained for derivatives **1** (*black*), **2** and **3** under identical conditions. **b** Schematic representation of the ΔG values measured for the binding of aminoglycosides **1**, **2**, **3**, **5** and **6** to the A-site

pH 7.0). After addition of the enzyme (0.2–1.0 μM), the evolution of the reaction mixture at 310 K was monitored by 1D-NMR, and the final products were characterized by 2D-NMR and MALDI-TOF MS. As shown in Fig. 9, at 100 μM neomycin-B, 100 μM ATP, 2.5 mM $MgCl_2$ and 0.2 μM ANT(4′), the adenylation reaction is completed within 22 min. For the rigid analogue **2**, under identical conditions, no reaction was detected even 24 h after enzyme addition (see Fig. 12). In contrast, the related deoxy derivative **5** reacted smoothly to give the monoadenylated compound within a few minutes, which demonstrates that the behavior exhibited by **2** has its origin in the extra rigidity of the antibiotic and that is not related to the absence of $OH5_{III}$. Therefore, the simple modification leading to conformationally restricted **2** provides an effective protection against aminoglycoside inactivation by *S. aureus* ANT(4′).

Surprisingly, the locked analogue **3**, exhibited a similar behavior to that observed for the natural antibiotic, neomycin-B. Indeed, the specific activities of ANT(4′) for aminoglycosides **1** and **3** are almost identical (see Fig. 12). According to these data, it seems that ANT(4′) is able to recognize and inactivate aminoglycoside conformations different to the anti-Ψ geometry observed in the enzyme-kanamycin complex [37] by X-ray methods. In fact, this conformation is not accessible for derivative **3**. In any case, these results prove that some degree of conformational flexibility in the ligand is critical for enzymatic activity. For ANT(4′) a strong conformational restriction (as present in **2**) prevents the antibiotic enzymatic adenylation, although the limited flexibility present in **3** restores the enzymatic activity.

In vivo activities of neomycin-B (**1**) and mimic **2** were tested employing the bacterium *E. coli* DH5α (pBBRIMCS-2), which expresses the resistance

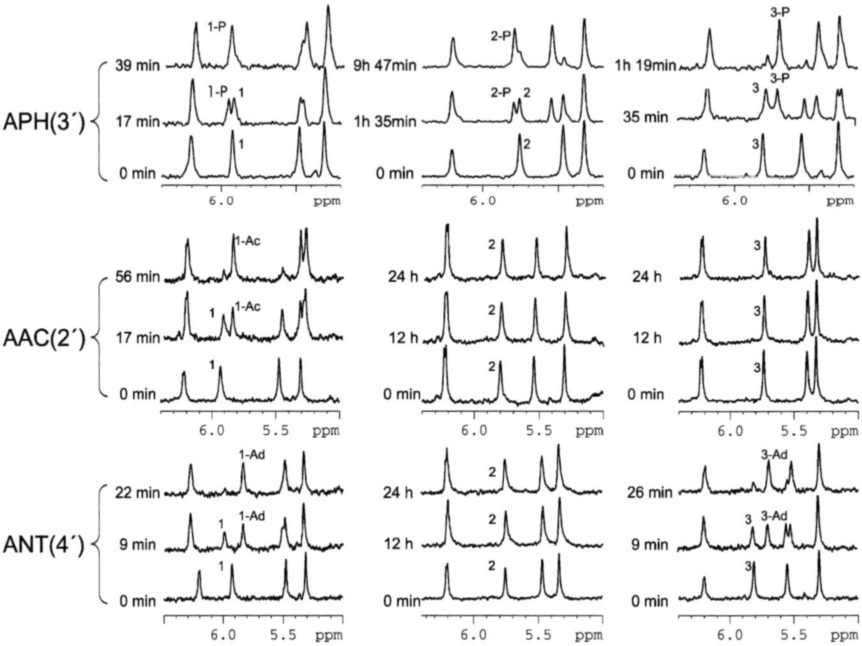

Fig. 12 Time evolution (at pH 7.0 and 310 K) of neomycin-B (**1**, *left*), derivative **2** (*middle*) and derivative **3** (*right*) at 100 μM concentration in the presence of ATP or AcCoA (100 μM), after addition of 0.2 μM ANT(4′) (*lower part*), AAC(2′) (*middle*) or APH(3′) (*upper part*) followed by 1-D NMR. **1-Ad** and **3-Ad** stand for the adenylated derivatives **1** and **3**, respectively. **1-Ac** stands for the acetylated derivative **1**. **1-P**, **2-P** and **3-P** stand for the phosphorylated derivatives **1**, **2** and **3**, respectively

enzyme ANT(4′). As expected, the MIC value for the natural antibiotic is significantly increased (from 3 to 60 μg/ml). In contrast, the cyclic derivative **2** maintains the same activity observed for the non-resistant bacteria (20 μg/ml). In conclusion, this simple modification leading to the conformationally restricted **2** provides an effective protection against aminoglycoside inactivation by *S. aureus* ANT(4′), both in vivo and in vitro, while maintaining a significant antibiotic activity.

Mycobacterium tuberculosis AAC(2′) catalyzes the transfer of one acetyl group from Acetyl CoA to position N-2 of the ring I. The 3D structure of AAC(2′) complexed to ribostamycin has been recently determined by X-Ray [38]. As previously mentioned, the oligosaccharide conformations bound by the RNA and by the enzyme are significantly different (see Fig. 7). Interestingly, the recognition of a high energy conformation of the antibiotic seems essential for enzymatic activity. Thus, by forcing an anti-Ψ geometry around the III/II linkage, the protein increases the accessibility of the reacting $NH2_I$ group which, otherwise, would be hydrogen-bonded to the non-vicinal unit III in the low energy syn-Ψ minimum. According to the conformational analysis described above, for cyclic derivatives **2** and **3**, this conformational change is no longer possible. Therefore, the geometrical restriction introduced in these derivatives is expected to fully protect the antibiotics from the enzymatic acetylation. In order to test this hypothesis, enzymatic reactions were performed in NMR tubes employing similar conditions to those previously described for ANT(4′). As shown in Fig. 12, at 100 μM neomycin-B, 100 μM AcCoA, 2.5 mM $MgCl_2$ and 0.2 μM AAC(2′), the acetylation reaction is completed within 1 h. A similar behavior was observed for the deoxy derivatives **5** and **6**. In contrast, no reaction was observed for the locked aminoglycosides **2** and **3**, even after 24 h and higher enzyme concentrations (0.2–20 μM). Therefore, the conformational constraint introduced in **2** and **3** provides complete protection against aminoglycoside modification by AAC(2′).

The in vivo activities of neomycin-B (**1**) and mimic **3** were tested employing the bacteria *E. coli* BL21(pET23a-AAC(2′)), which expresses the resistance enzyme AAC(2′). The MIC values measured for **1** and **3** were 2 and 2.5 μg/ml, respectively. As expected, **3** is not affected by the expression of AAC(2′) and this locked aminoglycoside exhibit a good antibiotic activity. However, under the employed experimental conditions, the AAC(2′) enzymatic activity seems to have a low influence on the antibiotic properties of the natural compound **1**. Although merely speculative, acetylation at position 2 of ring I might have a small influence on the stability of the drug/A-site complex. In fact, according to the structural information available [21–26] the amino group located at this position of the antibiotic is not involved in any key contact with the RNA target and therefore acetylated **1** might retain significant activity.

Finally, the effect of neomycin-B flexibility on its enzymatic inactivation by *Enterococcus faecalis* APH(3′) was tested. This enzyme catalyzes the transfer of a phosphate group from ATP to position 4 of ring I. In this case, the X-ray structure of the enzyme in complex with neomycin-B shows that the antibiotic structure recognized by the enzyme and the target RNA are remarkably similar (see Fig. 7) [40]. In fact, they mainly differ in the conformation adopted by the furanose ring of III (C3-endo and close to C2-endo for the RNA- and enzyme-bound conformations respectively).

As shown in Fig. 12, at 100 μM neomycin-B, 100 μM ATP, 2.5 mM $MgCl_2$ and 0.2 μM APH(3′), a 50% of antibiotic transformation is achieved in 17 min and the reaction is almost complete in 40 min. Under identical conditions, the enzyme needs 35 min to modify 50% of analogue **3**. For the more rigid derivative **2**, the reaction is even slower (around 100 min for a 50% transformation). In conclusion, although the extra rigidity of the locked derivatives provides a certain protection against modification by APH(3′), in this case, it is insufficient to prevent the enzymatic inactivation of the drugs. This result is not totally unexpected, given the close similarity of the aminoglycoside structures in the binding pockets of both receptors (the A-site RNA and the enzyme) [21–26, 40].

The in vivo activities of neomycin-B (**1**) and mimic **2** were tested employing the bacteria *E. coli* DH5α(pACYT177/ET3d/yNMT), which expresses the resistance enzyme APH(3′). The MIC values for the natural antibiotic and cyclic derivatives **2** and **3** were in all cases > 150 μg/ml. Therefore, in this case, although the extra rigidity present in compound **2** provides a certain in vitro protection against the antibiotic enzymatic inactivation, this is clearly insufficient to warrant good biological activity.

4 Conclusions

We have shown that the conformational behavior of natural aminoglycosides is characterized by a remarkable flexibility, with different conformations, even *non-exo-anomeric* ones, in fast exchange. This feature allows the adaptation of these ligands to the spatial and electronic requirements of the different receptors. The large diversity of structures adopted by aminoglycosides in the binding pocket of the different RNA receptors and the enzymes involved in bacterial resistance is consistent with this view.

We have also shown that the conformational differences exhibited by aminoglycosides within the binding pockets of the ribosome and of those enzymes involved in bacterial resistance can, in certain favorable cases, be exploited for designing new antibiotic derivatives not susceptible to enzymatic inactivation. Thus, the neomycin-B locked derivatives **2** and **3** have been specifically designed to mimic the bioactive RNA-bound structure of the drug. In fact, both antibiotics maintain a reasonable affinity for the target A-site RNA and exhibit significant biological activities. For compound **2**, the conformational constraint provides full protection against the aminoglycoside modification by *Staphylococcus aureus* ANT(4′) and *Mycobacterium tuberculosis* AAC(2′) and partially protects against modification by *Enterococcus faecalis* APH(3′). In fact, **2** presents an improved biological activity with respect to neomycin-B in bacteria expressing *Staphylococcus aureus* ANT(4′). In addition, derivative **3** is not susceptible to modification by *Mycobacterium tuberculosis* AAC(2′).

In our opinion, this example represents a test case for the validity of a structure-based approach for the design and preparation of ligands that specifically interact with a given receptor and that, in this particular case, might potentially be used as antibiotics. Finally, we would like to stress that the utility of conformational restriction to prevent the enzymatic inactivation of a particular drug while preserving its biological activity might not be restricted to aminoglycosides and could find further applications within medicinal chemistry.

References

1. Walter F, Vicens Q, Westhof E (1999) Curr Opin Chem Biol 3:694–704
2. Hermann T (200) Angew Chem Int Ed 39:1890–1905
3. Schroeder R, Waldsich C, Wank H (2000) EMBO J 19:1–9
4. Sucheck JS, Wong CH (2000) Curr Opin Chem Biol 4:678–686
5. Magnet S, Blanchard J (2005) Chem Rev 105:477–497
6. Smith CA, Baker EN (2002) Curr Drug Targets 2:143–160
7. Espinosa JF, Cañada J, Asensio JL, Dietrich H, Martín-Lomas M, Schmidt RR, Jiménez-Barbero J (1996) Angew Chem Int Ed Engl 35:303–306
8. Espinosa JF, Cañada J, Asensio JL, Dietrich H, Martín-Lomas M, Schmidt RR, Jiménez-Barbero J (1996) Angew Chem 108:323–326
9. Espinosa JF, Montero E, Vian A, Garcia J, Dietrich H, Martín-Lomas M, Schmidt RR, Imberty A, Cañada J, Jiménez-Barbero J (1998) J Am Chem Soc 120:10862–10871
10. Milton MJ, Bundle DR (1998) J Am Chem Soc 120:10547–10548
11. Garcia A, Montero E, Muñoz JL, Espinosa JF, Vian A, Asensio JL, Cañada FJ, Jimenez-Barbero J (2002) J Am Chem Soc 124:4804–4810
12. Dabrowski J, Kozar T, Grosskurth H, Nifantév NE (1995) J Am Chem Soc 117:5534–5539
13. Landersjo C, Stenutz R, Widmalm G (1997) J Am Chem Soc 119:8695–8698
14. Asensio JL, Hidalgo A, Cuesta I, González C, Cañada J, Vicent C, Chiara JL, Cuevas G, Jiménez-Barbero J (2002) J Chem Soc Chem Comm 2232–2233
15. Asensio JL, Hidalgo A, Cuesta I, González C, Cañada J, Vicent C, Chiara JL, Cuevas G, Jiménez-Barbero J (2002) Chem Eur J 8(22):5228–5240
16. Pearlman DA (1994) J Biomol NMR 4:1–16
17. Pearlman DA, Case DA, Caldwell JW, Ross WS, Cheatham TE, DeBolt S, Ferguson D, Siebal G, Kollman P (1995) Comp Phys Commun 91:1–41
18. Asensio JL, García A, Murillo MT, Fernández-Mayoralas A, Cañada FJ, Johnson CR, Jiménez-Barbero J (1999) J Am Chem Soc 121:11318–11331
19. Xu Q, Gitti R, Bush CA (1996) Glycobiology 6:281–288
20. Miller KE, Mukhopadhyay C, Cagas P, Bush CA (1992) Biochemistry 31:6703–6709
21. Carter AP, Clemons WM, Brodersen DE, Morgan-Warren RJ, Wimberly BT, Ramakrishnan V (2000) Nature 407:340–348
22. Vicens Q, Westhof E (2001) Structure 9:647–658
23. Vicens Q, Westhof E (2002) Chem Biol 9:747–755
24. Yoshizawa S, Fourmy D, Puglisi JD (1998) EMBO J 17(22):6437–6448
25. Fourmy D, Recht MI, Blanchard AC, Puglisi JD (1996) Science 274:1367–1371
26. Lynch SR, Ruben L, Gonzalez RL Jr, Puglisi JD (2003) Structure 11:43–53
27. Kondo S, Iinuma K, Yamamoto H, Maeda K, Umezawa H (1975) J Antibiot 26:412–417
28. Kondo S, Hotta K (1999) J Infect Chemother 5:1
29. Inoue M, Nonoyama M, Okamoto R, Ida T (1994) Drugs Exp Clin Res 20:233–239
30. Fujimura S, Tokue Y, Takahashi H, Nukiwa T, Hisamichi K, Mikami T, Watanabe A (1998) J Antimicrob Chemother 41:495–499
31. Roestamadji J, Graspas I, Mobashery S (1995) J Am Chem Soc 117:11060–11069
32. McKay GA, Roestamadji J, Mobashery S, Wright GD (1996) Antimicrob Agents Chemother 40:2648
33. Haddad J, Vakulenko S, Mobashery S (1999) J Am Chem Soc 121:11922–11923
34. Grapsas I, Lerner SA, Mobashery S (2001) Arch Pharm 334:295
35. Smith CA, Baker EN (2002) Curr Drug Targets 2:143–160
36. Vicens Q, Westhof E (2003) Biopolymers 70:42–57
37. Pedersen LC, Benning MM, Holden HM (1995) Biochemistry 34:13305–13311

38. Vetting MW, Hegde SS, Javid-Majd F, Blanchard JS, Roderick SL (2002) Nat Struct Biol 9:653–658
39. Vetting MW, Magnet S, Nieves E, Roderick SL, Blanchard JS (2004) Chem Biol 11:565–573
40. Fong DH, Berghuis AM (2002) EMBO J 21:2323–2331
41. Asensio JL, Hidalgo A, Bastida A, Torrado M, Corzana F, García-Junceda E, Cañada J, Chiara JL, Jiménez-Barbero J (2005) J Am Chem Soc 127:8278–8279
42. Kaul M, Barbieri CM, Pilch DS (2004) J Am Chem Soc 126:3447–3453
43. Shandrick S, Zhao Q, Han Q, Ayida BK, Takahashi M, Winters GC, Simonsen KB, Vourloumis D, Hermann T (2004) Angew Chem Int Ed 43:3177–3182
44. Wang Y, Hamasaki K, Rando RR (1997) Biochemistry 36:768–779
45. Ryu DH, Litovchick A, Rando RR (2002) Biochemistry 41:10499–10509
46. Fourmy D, Recht MI, Puglisi JD (1998) J Mol Biol 277:347–362
47. Blount KF, Zhao F, Hermann T, Tor Y (2005) J Am Chem Soc 127(27):9818–9829
48. Zhao F, Zhao Q, Blount KF, Han Q, Tor Y, Hermann T (2005) Angew Chem Int Ed 44:5329–5334

Top Curr Chem (2008) 273: 139–154
DOI 10.1007/128_2007_20

Published online: 29 February 2008

Solid State NMR Structure Analysis of the Antimicrobial Peptide Gramicidin S in Lipid Membranes: Concentration-Dependent Re-alignment and Self-Assembly as a β-Barrel

Sergii Afonin[1] · Ulrich H.N. Dürr[2] · Parvesh Wadhwani[1] · Jesus Salgado[3] · Anne S. Ulrich[1,4] (✉)

[1]Karlsruhe Institute of Technology (KIT), Institut für Biologische Grenzflächen, PO Box 3640, 76021 Karlsruhe, Germany
anne.ulrich@ibg.fzk.de

[2]Max Planck Institut für Biophysikalische Chemie, Department of NMR-Based Structural Biology, Fassberg 11, 37077 Göttingen, Germany

[3]Instituto de Sciencia Molecular, Universitat de València, Poligono la Coma s/n, 46980 Paterna (Valencia), Spain

[4]Karlsruhe Institute of Technology (KIT), Institut für Organische Chemie, Fritz-Haber-Weg 6, 76131 Karlsruhe, Germany

Abstract Antimicrobial peptides can kill bacteria by permeabilizing their cell membrane, as these amphiphilic molecules interact favourably with lipid bilayers. This mechanism of action is attributed either to the formation of a peptide "carpet" on the membrane surface, or to a transmembrane pore. However, the structure of such a pore has not yet been resolved under relevant conditions. Gramicidin S is a symmetrical cyclic β-sheet decapeptide, which has been previously shown by solid state NMR to lie flat on the membrane surface at low peptide:lipid ratios (≤1 :80). Using highly sensitive ^{19}F-NMR, supported by ^{15}N-labelling, we found that gramicidin S can flip into an upright transmembrane alignment at high peptide:lipid ratios (≥1 :40). Orientational NMR constraints suggest that the peptide may self-assemble as an oligomeric β-barrel pore, which is stabilized by intermolecular hydrogen bonds. Comparison of different model membranes shows that the observed re-alignment is favoured in thin bilayers with short-chain lipids, especially near the chain melting temperature, whereas long-chain lipids suppress pore formation. Based on the oligomeric structural model and the conditions of pore formation, guidelines may now be derived for rationally designing peptide analogues as antibiotics with improved selectivity and reduced side effects.

Keywords Amphiphilic peptide–lipid interactions · Cyclic β-sheet structure · Peptide re-alignment and self-assembly · Pore formation · Solid state ^{19}F-NMR and ^{15}N-NMR

Abbreviations

DPPC	1,2-Dipalmitoyl-*sn*-glycero-3-phosphocholine
DLPC	1,2-Dilauroyl-*sn*-glycero-3-phosphocholine
DMPC	1,2-Dimyristoyl-*sn*-glycero-3-phosphocholine
GS	Gramicidin S
GS-1/1′	Gramicidin S labelled with 4F-Phg at Val1/Val1′ and with ^{15}N-Leu in 3/3′
GS-3/3′	Gramicidin S labelled with 4F-Phg at Leu3/Leu3′ and with ^{15}N-Val in 1/1′
4F-Phg	4-Fluoro-phenylglycine
τ	Tilt angle of the peptide symmetry axis with respect to the membrane normal
ρ	Azimuthal rotation angle of the peptide
S_{mol}	Molecular order parameter of the peptide

1
Introduction

Increasing resistance of bacterial strains against conventional antibiotics calls for new types of drugs to combat infection. Cationic antibiotic peptides offer a promising potential as they supposedly disrupt bacterial membranes mechanically in a receptor-independent way [1, 2]. Hundreds of such peptides are known from diverse organisms, ranging from bacteria to humans. One of the smallest representatives is gramicidin S (GS), cyclo-(Val-Orn-Leu-^{D}Phe-Pro)$_2$ from *Aneurinibacillus migulanus* (*Bacillus brevis*), consisting of a cyclic antiparallel β-sheet structure that is stabilized by four intramolecular hydrogen bonds (see Fig. 1A) [3, 4]. It has a pronounced amphiphilic structure, which is characteristic of antimicrobial peptides and essential for their general function.

Several different modes of membrane disruption have been proposed as possible mechanisms of action for peptide antibiotics in general [2, 5]:

i A "carpet" of peptides at the membrane surface could lead to bilayer rupture
ii Several peptide molecules could immerse into the membrane to form a pore
iii The pore may additionally contain negatively charged lipid headgroups
iv The amphiphilic peptides may exert a detergent-like action

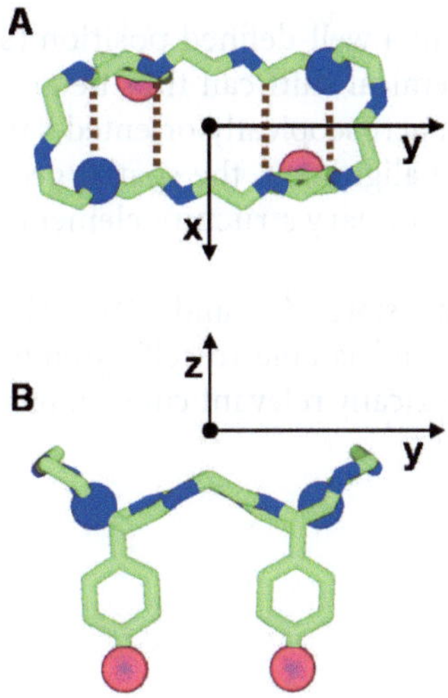

Fig. 1 Chemical structure of the gramicidin S analogue GS-3/3′ with 4F-Phg (Fluorine atom is marked *red*) substituted for Leu3/Leu3′ and ^{15}N-labeled Val1/Val1′ (Nitrogen atom is marked *blue*). The peptide is displayed from one hydrophilic surface (**A**) and from one side (**B**) to define the molecular axis system used in the structure analysis

Conductivity measurements have demonstrated the pore-forming ability of gramicidin S and other antimicrobial peptides [6], but a corresponding three-dimensional structure has not yet been resolved under relevant conditions. It should nevertheless be possible to detect such an arrangement in membranes by measuring the alignment of the peptides with respect to the lipid bilayer. Solid-state nuclear magnetic resonance (NMR) can provide the necessary orientational constraints of membrane-bound peptides under quasi-native conditions and is also sensitive to dynamic processes. As an NMR strategy, orientational constraints are extracted from selectively isotope-labelled peptides, which are embedded in membranes that are macroscopically oriented and aligned horizontally in the static magnetic field [7, 8]. For example, selective or uniform ^{15}N-labels in the backbone are readily used to reconstruct the overall conformation and alignment of a peptide or protein in a membrane [9, 10], although they suffer from low sensitivity. Additionally, local distances can be obtained from the dipolar coupling between labels, either by static methods [11, 12] or by magic angle spinning [13–15].

The generally low sensitivity of solid state NMR is significantly enhanced by using selective ^{19}F-labels to analyze the structures of membrane-associated peptides and proteins [13, 16–24]. This opens up a previously inaccessible range of experimental conditions, such as peptide:lipid ratios as low as 1:3000 [25]. This is particularly relevant for antibiotic peptides, as their concentration dependence constitutes an essential functional aspect [26]. A suitable side chain for labelling is 4F-phenylglycine (4F-Phg), as the ^{19}F-reporter is fixed to the peptide backbone in a well-defined position (see Fig. 1B) [27,28]. The anisotropic ^{19}F-NMR chemical shift can thus be analyzed to obtain orientational constraints from macroscopically oriented samples. These are used to determine the membrane alignment, the spatial geometry, and the dynamic behaviour of the entire secondary structure element to which the ^{19}F-label is attached.

Here we show by solid state ^{19}F- and ^{15}N-NMR that gramicidin S can re-align in lipid bilayers and is able to self-assemble as a trans-membrane β-barrel pore under biologically relevant conditions.

2 Materials and Methods

2.1 Synthesis of Labelled Peptides

Linear peptides were synthesized on an Applied Biosystems 433A synthesizer with standard solid-phase Fmoc protocols. Cyclization and side-chain deprotection were performed as previously described [27,29] and monitored by reverse phase HPLC (Jasco, Japan; 4.6 × 250 mm Vydac C18 column). Crude peptides were purified with water/acetonitrile gradients, using 5 mM HCl to avoid and remove trifluoroacetic acid, and peptide mass was confirmed by MALDI-TOF. Given the strong tendency of phenylglycine to undergo racemization during peptide synthesis, a racemic mixture of 4F-Phg was employed. Epimeric peptides were then separated by HPLC and identified by derivatizing the hydrolyzates with Marfey's reagent [27].

2.2 Preparation of Oriented NMR Samples

The lipids 1,2-dimyristoyl-*sn*-glycero-3-phosphocholine (DMPC), 1,2-dipalmitoyl-*sn*-glycero-3-phosphocholine (DPPC), and 1,2-dilauroyl-*sn*-glycero-3-phosphocholine (DLPC), were obtained from Avanti Polar Lipids (Alabaster, AL) and used without further purification. Three types of samples were investigated:

i Pure peptides in the form of a lyophilized powder
ii Lipid–peptide suspensions, in which the peptide was mixed and extruded to give large unilamellar liposomes in excess (>50% w/w) aqueous buffer
iii Reconstituted lipid bilayers that were macroscopically oriented on a solid support

Oriented membrane samples were prepared by depositing peptide–lipid mixtures (co-solubilized in $CHCl_3$/MeOH 2:1, v/v) on glass slides ($0.08 \times 7.5 \times 18\,mm^3$, Paul Marienfeld KG, Mergentheim, Germany). Drops of 20–30 μL containing up to 1.2 mg material (per slide) were spread onto each glass plate and dried under vacuum overnight. The plates were stacked and hydrated for 2–4 days in an atmosphere of saturated K_2SO_4 solution at a temperature above the lipid phase transition. The stack was wrapped in parafilm and polyethylene foil to maintain hydration during the NMR measurement. Likewise, samples (i) and (ii) were sealed in small plastic bags.

2.3 Solid State ^{19}F-, ^{15}N-, and ^{31}P-NMR Measurements

NMR measurements were performed on 500 MHz Unity Inova (Varian, Palo Alto, CA) and Avance (Bruker Biospin, Karlsruhe, Germany) widebore spectrometers, operating at 470.2 MHz for ^{19}F, 202.4 MHz for ^{31}P, and 50.68 MHz for ^{15}N. For solid state ^{19}F-NMR, a ^{19}F/^{1}H double-tuned flat-coil probe (Doty Scientific, Columbia, SC) and an external 470 MHz high power amplifier (Creative Electronics, Los Angeles, CA) were used. ^{15}N- and ^{31}P-NMR experiments were performed on a ^{1}H/X/Y Bruker probe. Both probes are equipped with a susceptibility-matched housing that allows manual rotation of the oriented samples with respect to the external magnetic field. For ^{19}F- and ^{31}P-NMR continuous-wave ^{1}H-decoupled Hahn-echo spectra of oriented samples were recorded with 90° pulse durations of 2.2 μs (^{19}F) and 6 μs (^{31}P), relaxation delays of 2.4 s (^{19}F) and 3 s (^{31}P), and inter-pulse echo delay times of typically 25 μs. ^{15}N was detected via a ramped cross-polarization (CP) sequence with a CP power of 40 kHz, 2.5 s relaxation delay time, 100 kHz spectral width, 2048 data points, and tppm20 proton decoupling. The CP contact time was optimized for each type of sample and varied between 2.5 ms (static powders) and 100–200 μs (oriented samples). The ^{19}F chemical shifts were referenced to $CFCl_3$ by setting the signal of 100 mM NaF (in 10 mM HEPES, pH 7.5, measured at 35 °C) to −120.4 ppm. For ^{31}P direct referencing with a primary standard (85% H_3PO_4 solution, assigned to 0 ppm) was used. ^{15}N-NMR spectra were referenced relative to $^{15}NH_4NO_3$ by setting the signal of solid $^{15}NH_4Cl$ to 18.0 ppm. The static ^{19}F and ^{15}N CSA tensor elements and frequencies of broad resonance lines were determined by fitting the spectra with a self-written C++ program and DMfit [30].

3 Results and Discussion

3.1 ^{19}F-NMR of Gramicidin S in Oriented Membranes

We have demonstrated for many different peptides that substitution of a single hydrophobic amino acid by L-4F-Phg does not spoil their conformation or biological function [18, 19, 21, 31]. Gramicidin S analogues carrying two ^{19}F-labels in place of either Leu3/Leu3′ or Val1/Val1′ exhibit antimicrobial activities comparable to the wild type [27]. Their structural integrity was confirmed by solid state NMR from the dipolar couplings within each pair of symmetrically apposed labels. The previously reported ^{19}F–^{19}F distance of 6 Å [17] in the membrane-bound Leu-3/3′-substituted peptide analogue (GS-3/3′) agrees well with the wild-type conformation in dimethyl sulfoxide [4]. Likewise, Val-substituted, GS-1/1′, reveals a distance of 8 Å in the membrane, as expected [22].

To describe the alignment and motional averaging of gramicidin S in a membrane, we use three parameters that can be calculated from orientational constraints [19, 22]. Namely, the tilt angle τ defines the angle between the membrane normal and the main peptide symmetry axis (z-axis in Fig. 1), the azimuthal rotation ρ describes the alignment of the β-strands (rotation around z-axis), and the molecular order parameter S_{mol} quantifies the degree of motional averaging on a scale from 1 (no wobble) to 0 (free isotropic tumbling). We have previously analyzed GS-3/3′ in fully hydrated DMPC bilayers at low peptide:lipid ratios of $\leq 1{:}80$ (mol/mol) [17]. A representative ^{19}F-NMR spectrum is illustrated in Fig. 2A, showing a narrow resonance at −119 ppm. The presence of a single signal indicates that both equivalent

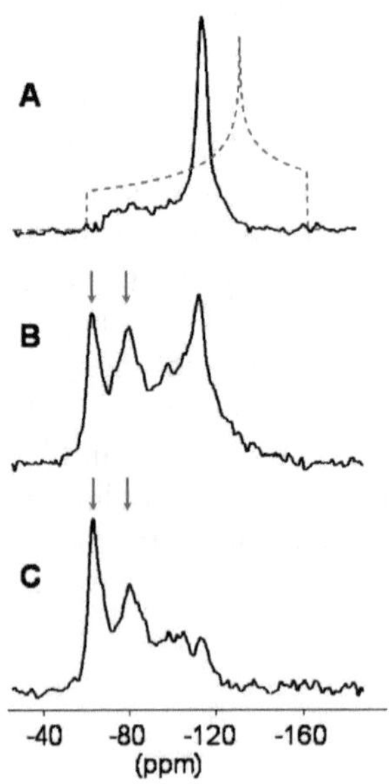

Fig. 2 Solid state ^{19}F-NMR spectra of GS-3/3′ in oriented DMPC bilayers, measured at 25 °C according to published procedures [17–19, 22]. Samples are prepared with a peptide:lipid molar ratio of 1:80 (**A**), 1:40 (**B**), and 1:20 (**C**). The *arrows* indicate the signals emerging at high peptide concentration, corresponding to a new re-aligned state of the peptide. The *dashed line* shows respective powder pattern of the lyophilized peptide

4F-Phg labels are aligned the same way with respect to the membrane normal. This observation, together with a comprehensive mathematical data analysis, shows that at low concentration the peptide assumes a flat orientation in the membrane ($\tau = 0°$, ρ unconstrained due to symmetry), as expected from its amphiphilic structure [17]. In liquid crystalline DMPC (temperature >23 °C) gramicidin S has a small order parameter of $S_{mol} \approx 0.3$, suggesting that it is highly mobile and thus monomeric. In the gel state (<23 °C) the peptide is immobilized ($S_{mol} \approx 1$) but retains its flat surface alignment. A more recent analysis of the differently labelled GS-1/1′ analogue confirms this molecular behaviour (data not shown).

Here, we have analyzed gramicidin S at higher peptide:lipid ratios than before, namely at 1:40 and 1:20 [32, 33]. These concentrations are biologically relevant in view of the high limiting threshold concentration that needs to be exceeded for bacterial cell lysis [34, 35]. Remarkably, a new set of signals emerges in the ^{19}F-NMR spectra of GS-3/3′ at –65 and –80 ppm $\pm$ 3 ppm, as indicated by the arrows in Fig. 2. In the case of GS-1/1′ a new signal emerges at –86 ppm, representing two overlapping resonances that are unresolved (data not shown). These new signals show that the peptide must assume a differ-

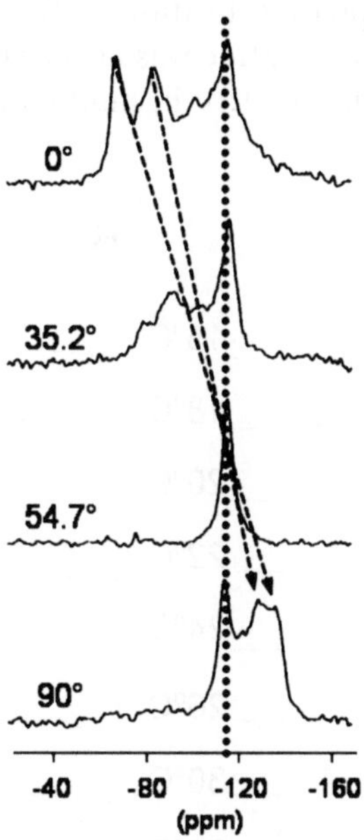

Fig. 3 ^{19}F-NMR spectra at different sample orientations. The oriented sample of Fig. 2B was manually tilted in the static magnetic field B_0 to acquire spectra at different angles, where $\theta = 0°$ corresponds to the usual alignment of the sample normal parallel to the direction of B_0. The *arrows* indicate the shift of the anisotropic signals with respect to the isotropic value (*dotted line*), according to the factor $3(\cos^2\theta - 1)/2$ as expected for a molecule undergoing fast long-axial rotation about the membrane normal

ent orientation in the membrane compared to the flat surface-bound state that was described at low peptide concentration (cf. Fig. 2A). Virtually complete re-alignment of gramicidin S is seen in Fig. 2C at a high peptide:lipid ratio of 1:20. Figure 2B illustrates an intermediate situation at 1:40, where only half the peptides have flipped into the new orientational state. The observation of two distinct new NMR signals for GS-3/3′ suggests that the two 4F-Phg segments in positions 3 and 3′ have become spectroscopically inequivalent due to different individual orientations in the membrane. The occurrence of a ^{19}F-NMR signal at the very edge of the static powder lineshape of 4F-Phg (dashed line in Fig. 2A) indicates that the peptide does not wobble in the membrane ($S_{\mathrm{mol}} \approx 1$). It is nevertheless engaged in fast rotation around the membrane normal, because the spectral lineshape is scaled by a factor of $-1/2$ when the oriented sample is measured vertically in the magnetic field, as seen in Fig. 3.

3.2 Examination of the Phospholipids by ^{31}P-NMR

The quality of alignment in the oriented membrane samples was examined by ^{31}P-NMR. Figure 4A shows that the lipids are well oriented at all temperatures above and below the lipid phase transition at $T_{\mathrm{m}} \approx 23\,^{\circ}$C. Any potential disturbance of the bilayer by the peptide was also examined by ^{31}P-NMR, using a very high peptide:lipid ratio of 1:10 in multilamellar vesicles. The charac-

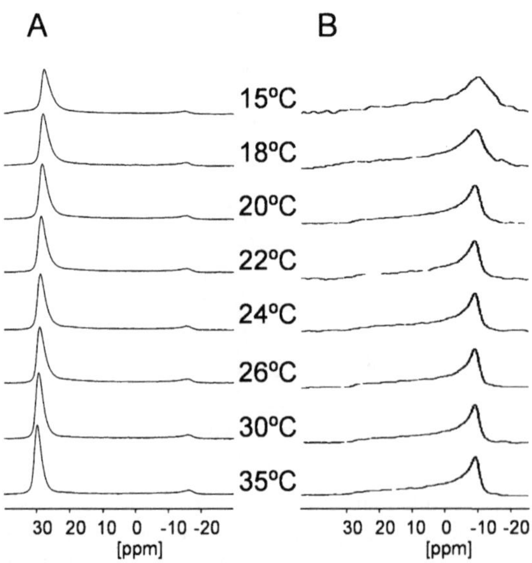

Fig. 4 ^{31}P-NMR spectra of phospholipids reconstituted with GS, showing **A** the same oriented sample as in Fig. 2A (1:20), and **B** dispersion of unilamellar DMPC vesicles in excess buffer containing gramicidin S in a 1:10 peptide:lipid ratio

teristic lineshapes of Fig. 4B demonstrate that the lamellar bilayer structure is maintained without defects. There is no sign of isotropic or hexagonal signals, not even at the lipid phase transition temperature, hence gramicidin S is unlikely to induce a toroidal wormhole in the membrane, as had been suggested for some other antimicrobial peptides [10, 36–38].

3.3 Calculation of the Peptide Alignment

The peptide alignment is calculated by a χ^2-analysis of the ^{19}F-NMR chemical shifts as a function of τ and ρ (using $S_{mol} = 1$ as there is no wobble), as previously explained [18, 19, 22, 23]. Figure 5A,B shows the allowed regions in the $\tau-\rho$ maps for GS-3/3′ and GS-1/1′, respectively. The overlap of the allowed regions suggests a peptide tilt angle τ between 70° and 90°, but the value of the azimuthal rotation ρ is not well defined.

To analyze the 4F-Phg labels it was necessary to know their side chain torsion angle χ_1 around Cα–Cβ, in order to fix the ^{19}F chemical shift anisotropy tensor unambiguously with respect to the peptide backbone [17, 39]. The allowed range for χ_1 was thus determined by calculating the respective energy profiles for 4F-Phg in the intact GS-3/3′ and GS-1/1′ molecules, using the COSMOS force field [40]. Figure 6 illustrates that, depending on its substituted position, 4F-Phg experiences different steric interference from the diagonally apposed side chains of Val1/Val1′ or Leu3/Leu3′, respectively. Energy profiles were thus calculated for all side chain conformers of Val (gauche⁻, *trans*, gauche⁺) in the case of GS-3/3′ (Fig. 6A), and of Leu (tg⁻, tt, tg⁺, g⁻t) in the case of GS-1/1′ (Fig. 6B).

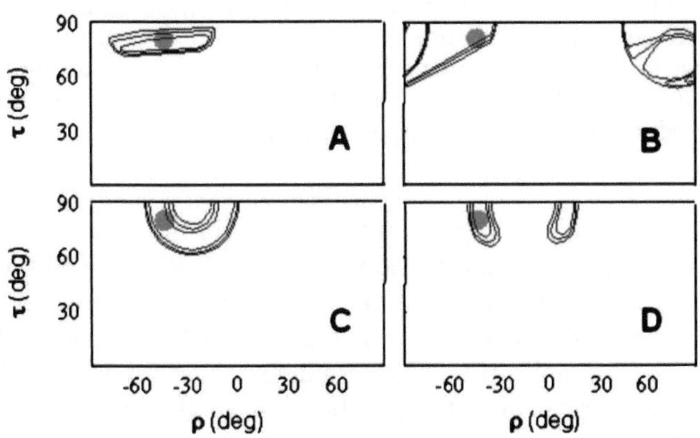

Fig. 5 Contour plots showing the alignment of gramicidin S in DMPC membranes at high peptide:lipid ratios (≥1 :40), evaluated in terms of the peptide tilt angle τ and its azimuthal rotation ρ. Feasible solutions from ^{19}F-NMR of GS-3/3′ (**A**) and GS-1/1′ (**B**) are shown, together with additional ^{15}N-NMR constraints from GS-3/3′ (**C**) and GS-1/1′ (**D**). A unique overlap is obtained at $\tau \approx 80°$ and $\rho \approx -45°$, as indicated by the *shaded dots*

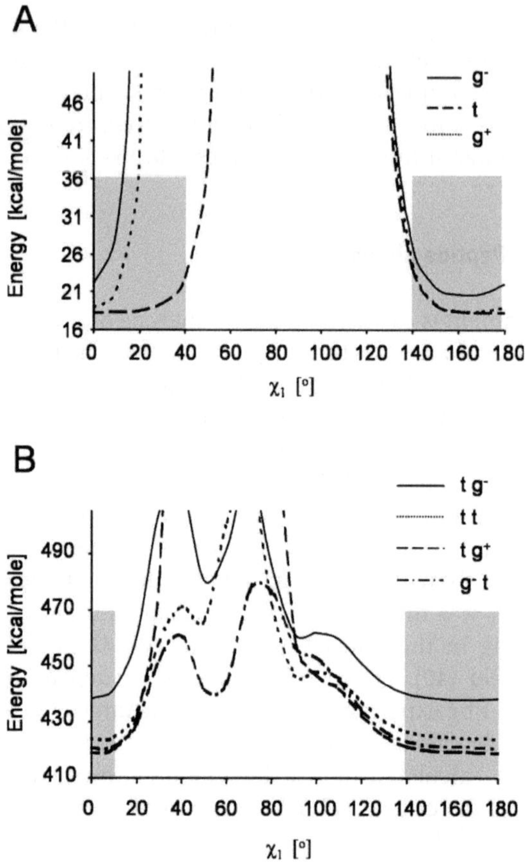

Fig. 6 Determination of the 4F-Phg side chain torsion angle χ_1 from the respective energy profiles for 4F-Phg in the intact molecules GS-3/3′ (**A**) and GS-1/1′ (**B**). An acceptable range for the 4F-Phg torsion angle is between $-40° < \chi_1 < 40°$ for GS-3/3′, and $-40° < \chi_1 < 10°$ for GS-1/1′ (*shaded regions*). These respective ranges were used to calculate the allowed regions in the τ/ρ-maps of Fig. 5A,B

Given that a certain ambiguity is introduced in the data analysis by having to estimate the side chain torsion angles of 4F-Phg, the allowed regions in the τ–ρ maps for GS-3/3′ and GS-1/1′ are not very precise. The value of ρ remains rather ill-defined from the ^{19}F-NMR data alone. Therefore, additional orientational constraints were determined from ^{15}N-Val and ^{15}N-Leu labels incorporated into the backbone of GS-3/3′ and GS-1/1′, respectively [29]. The experiments and spectral analysis were performed according to established solid state NMR procedures [7–10, 41]. ^{15}N-NMR spectra were acquired on the same oriented samples that had been used for ^{19}F-NMR. Just like Fig. 2B, Fig. 7 also shows two groups of signals corresponding to the two peptide populations at a peptide:lipid ratio of 1:40. The ^{15}N chem-

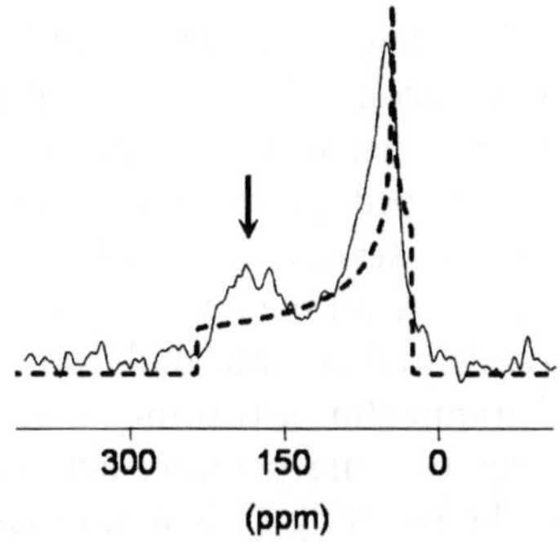

Fig. 7 Solid state ^{15}N-NMR spectra of GS-3/3′ at 1:40 in oriented DMPC bilayers. The *arrow* indicates the new ^{15}N resonance emerging at high peptide concentration, from which the upright alignment of gramicidin S in the membrane was calculated. The *dashed line* corresponds to the respective powder pattern of the lyophilized peptide

ical shift of 50 ppm ± 20 ppm for GS-3/3′ (and 60 ppm for GS-1/1′, data not shown) represents the flat surface alignment of gramicidin S, which predominates at low concentration. The other signal at 185 ppm ± 20 ppm for GS-3/3′ (and 155 ppm for GS-1/1′) represents the new peptide population that has flipped upright. The individual resonances of the two inequivalent ^{15}N labels in positions 1 and 1′ (or 3 and 3′, respectively) are not resolved in the flipped structure.

Based on the ^{15}N-NMR data alone, the corresponding areas in the $\tau-\rho$ maps are displayed in Fig. 5C (^{15}N-Val in GS-3/3′) and Fig. 5D (^{15}N-Leu in GS-1/1′). The overlap of these regions with the constraints from ^{19}F-NMR (Fig. 5A,B) thus defines the novel orientation of gramicidin S in membranes at high peptide concentration. The allowed area (common to all four panels) is marked by a dot in Fig. 5 at $\tau = 80° \pm 10°$ and $\rho = -45° \pm 10$. As the value of τ is close to a right angle between the molecular symmetry axis (*z*-axis, see Fig. 1) and the membrane normal, this means that the β-sheet plane (*x*–*y* plane) is tilted almost perpendicular with respect to the lipid bilayer plane. The value of ρ indicates that the β-strands (peptide *y*-axis) are inclined by about 45° with respect to the membrane normal.

3.4 Interpretation of the Re-aligned Peptide Structure

The flipped alignment of gramicidin S is unlikely to represent a monomeric peptide immersed in the membrane, as both its hydrophilic and hydrophobic surfaces would be laterally exposed in the lipid bilayer. However, there are two arguments suggesting that oligomerization may take place at high peptide concentration. First of all, we found that the re-aligned peptide undergoes virtually no wobble, since the order parameter is $S_{mol} \approx 1$. This is an unambiguous statement, since one of the signals in Fig. 2B,C lies at the very edge of the static CSA tensor. Second, the re-aligned peptide is engaged in fast rotation about the membrane normal, as evident from Fig. 3. Therefore, it is very likely that a number of flipped gramicidin S peptides are self-assembled as an oligomer. Indeed, the observed slant of the β-strands ($\rho \approx -45°$) is close to the typical slant in β-barrel membrane proteins [42, 43]. Such slant ideally disposes the β-strands to form intermolecular hydrogen bonds to either side. In the case of gramicidin S, two intermolecular hydrogen bonds would form with each neighbouring peptide, between Orn2NH→Pro5′CO and Orn2′NH→Pro5CO. The resulting ribbon of cross-β-strands will intrinsically acquire a right-handed twist and close up on itself as a ring, with the hydrophilic surface of GS facing inside. We performed simple computer docking experiments, which confirmed that the proposed peptide oligomer is sterically feasible, for example as a hexamer, with regard to hydrogen bond angles and distances. Figure 8 illustrates how the resulting β-barrel would look, though the number of monomers still remains speculative.

Amongst the multitude of three-dimensional structures of gramicidin S in the literature [4, 32], one asymmetric furanoid sugar-containing analogue indeed forms hexamers with similar intermolecular contacts in the crystal as were derived here in a quasi-native membrane environment [44]. Most

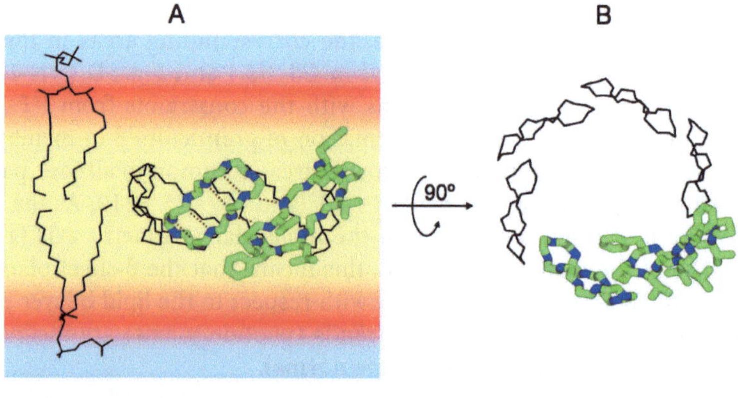

Fig. 8 Proposed model for gramicidin S in a membrane according to the orientational constraints obtained from ^{19}F- and ^{15}N-NMR. The upright backbone alignment ($\tau \approx 80°$) and slant of the β-sheets ($\rho \approx -45°$) are compatible with the formation of an oligomeric β-barrel that is stabilized by hydrogen bonds (*dotted lines*). **A** The oligomer is depicted sideways from within the lipid bilayer interior (showing only backbone atoms for clarity, but with hydrophobic side chains added to one of the monomers). Atomic coordinates of GS were taken from a monomeric structure [4], and the two DMPC lipid molecules are drawn to scale (from a molecular dynamics simulation coordinate file). The bilayer cross-section is coloured *yellow* in its hydrophobic core, *red* in the amphiphilic regions, and *light blue* near the aqueous surface. **B** Illustrates a top view of the putative pore, although the number of monomers remains speculative

striking is the resemblance between our model structure and the multi-stranded β-barrels known for various membrane proteins [42] and pore-forming toxins [43]. The formation of an aqueous pore in the lipid bilayer would indeed offer an explanation for the observed bilayer conductivity induced by gramicidin S upon membrane binding [6]. The peptide:lipid ratio of 1:40 at which this structure could be trapped for NMR analysis appears to be biologically relevant, as the minimum inhibitory concentration of gramicidin S corresponds to far more than an equimolar ratio of peptides per lipid molecule on the bacterial surface [34, 35].

3.5
Influence of the Bilayer Thickness

To find out which conditions favor the self-assembly and pore formation of gramicidin S, we systematically varied the peptide concentration, lipid environment, temperature, etc. The unique sensitivity of ^{19}F-NMR allowed spectra to be acquired within 2 h using only 0.4 mg of peptide. The spectra illustrated in Fig. 9 display the same characteristic features of the well-resolved data of Fig. 2, despite the lower number of scans. Figure 9A shows that at a peptide:lipid ratio of 1:40 in short-chain DLPC (containing 12 carbons in the acyl chain; $T_m \approx -1\,°C$) virtually all peptides can re-align and self-assemble over a wide temperature range of about 30 °C. By comparison, in DMPC (14 carbons, Fig. 9B) re-alignment occurs only within a range of

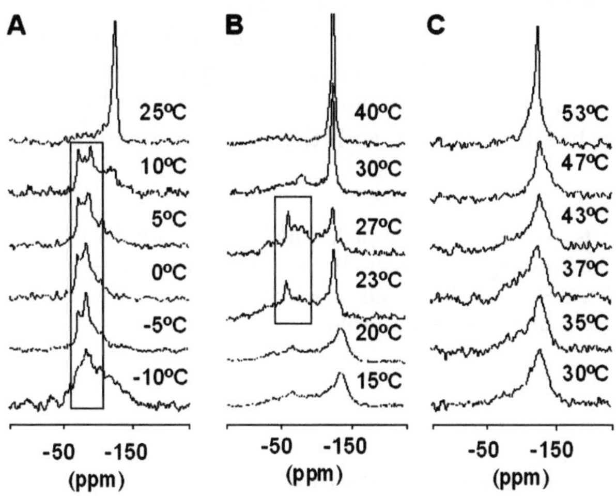

Fig. 9 ^{19}F-NMR spectra of 0.4 mg GS-3/3′ each, reconstituted at a peptide:lipid ratio of 1:40 in short acyl chain-length DLPC bilayers (**A**), medium chain-length DMPC (**B**), and long-chain DPPC (**C**). Each temperature series encompasses the phase transition of the respective lipid. Resonances corresponding to the new upright peptide alignment are highlighted by *boxes*

about 10 °C centred around $T_m \approx 23$ °C and involves no more than half the peptides. Finally, in long-chain DPPC (16 carbons, $T_m \approx 41$ °C, Fig. 9C) the peptide remains flat on the surface throughout all temperatures. Summarizing these results, it is found that GS favorably immerses into thin membranes at temperatures close to the lipid chain melting transition T_m. These findings suggest that the height of the GS β-barrel (approximately 18 Å) has to match the hydrophobic thickness of the membrane [22, 32]. It is interesting to note that living cells maintain their membranes in a phase state that is neither gel nor totally fluid [45]. Any phase boundaries, as well as membrane-spanning proteins, are thus likely to induce local bilayer defects at which GS may favorably insert and re-align.

4 Conclusions

We conclude that gramicidin S can assume two distinct states in model membranes: monomeric peptides are aligned flat on the membrane surface, while oligomers are assembled as a transmembrane β-barrel. These two states might correspond to an inactive form at low concentration and a membrane-disruptive pore above a limiting threshold concentration. Alternatively, it is also conceivable that the different structures may be correlated with different effects of GS on living membranes, namely its antimicrobial action against bacteria and its lytic effects on eukaryotic cells [46]. These questions will have to be addressed by biological activity assays using specifically modified peptide analogues, having recognized the possibility of self-assembly and the necessity for hydrogen bonding in the oligomeric state. It should thus be possible to rationally design new gramicidin S analogues with a modified hydrogen bonding pattern, with the aim of improving their therapeutic index.

Acknowledgements All authors acknowledge the Friedrich-Schiller-University of Jena and the Deutsche Forschungsgemeinschaft, as the project was started within SFB 197 and SFB 604. We thank Stephan Grage and Ulrich Sternberg for helpful discussions, and JS thanks the EMBO foundation for a stipend.

References

1. Hancock RE (2001) Lancet Infect Dis 1:156
2. Zasloff M (2002) Nature 415:389
3. Hull SE, Karlsson R, Main P, Woolfson MM, Dodoson EJ (1978) Nature 275:206
4. Xu Y, Sugár IP, Krishna R (1995) J Biomol NMR 5:37
5. McElhaney RN, Prenner EJ (1999) Biochim Biophys Acta 1462:1
6. Wu M, Maier E, Benz R, Hancock RE (1999) Biochemistry 38:7235

7. Bechinger B, Sizun C (2003) Concepts Magn Reson 18A:130
8. Strandberg E, Ulrich AS (2004) Concepts Magn Reson 23A:89
9. Opella SJ, Marassi FM (2004) Chem Rev 104:3587
10. Henzler-Wildman KA, Lee DK, Ramamoorthy A (2003) Biochemistry 42:6545
11. Grage SL, Ulrich AS (1999) J Magn Reson 138:98
12. Grage SL, Ulrich AS (2000) J Magn Reson 146:81
13. Toke O, O'Connor RD, Weldeghiorghis TK, Maloy WL, Glaser RW, Ulrich AS, Schaefer J (2004) Biophys J 87:675
14. Crocker E, Patel AB, Eilers M, Jayaraman S, Getmanova E, Reeves PJ, Ziliox M, Khorana HG, Sheves M, Smith SO (2004) J Biomol NMR 29:11
15. Porcelli F, Buck B, Lee DK, Hallock KJ, Ramamoorthy A, Veglia G (2004) J Biol Chem 279:45815
16. Ulrich AS (2000) High resolution solid state NMR, ^{1}H, ^{19}F. In: Lindon J, Tranter G, Holmes J (eds) Encyclopedia of spectroscopy and spectrometry. Academic, London, p 813
17. Salgado J, Grage SL, Kondejewski LH, Hodges RS, McElhaney RN, Ulrich AS (2001) J Biomol NMR 21:191
18. Afonin S, Dürr UHN, Glaser RW, Ulrich AS (2004) Magn Reson Chem 42:195
19. Glaser RW, Sachse C, Dürr UHN, Wadhwani P, Ulrich AS (2004) J Magn Reson 168:153
20. Getmanova E, Patel AB, Klein-Seetharaman J, Loewen MC, Reeves PJ, Friedman N, Sheves M, Smith SO, Khorana HG (2004) Biochemistry 43:1126
21. Glaser RW, Sachse C, Durr UH, Wadhwani P, Afonin S, Strandberg E, Ulrich AS (2005) Biophys J 88:3392
22. Ulrich AS (2005) Prog NMR Spectr 46:1
23. Ulrich AS, Wadhwani P, Dürr UHN, Afonin S, Glaser RW, Strandberg E, Tremouilhac P, Sachse C, Berditchevskaia M, Grage S (2006) Solid-state ^{19}F-nuclear magnetic resonance analysis of membrane-active peptides. In: Ramamoorthy A (ed) NMR spectroscopy of biological solids. Taylor & Francis, London, p 215
24. Ulrich AS (2007) Solid state ^{19}F-NMR analysis of oriented biomembranes. In: Webb GA (ed) Modern magnetic resonance. Springer, p 257
25. Glaser RW, Ulrich AS (2003) J Magn Reson 164:104
26. Huang HW, Chen FY, Lee MT (2004) Phys Rev Lett 92:198304
27. Afonin S, Glaser RW, Berditchevskaia M, Wadhwani P, Gührs KH, Möllmann U, Perner A, Ulrich AS (2003) ChemBioChem 4:1151
28. Mikhailiuk PK, Afonin S, Chernega AN, Rusanov EB, Platonov M, Dubinia G, Ulrich AS, Komarov IV (2006) Angew Chem Int Ed Engl 45:5659
29. Wadhwani P, Afonin S, Ieronimo M, Buerck J, Ulrich AS (2006) J Org Chem 71:55
30. Massiot D, Fayon F, Capron M, King I, Le Calve S, Alonso B, Durand J-O, Bujoli B, Gan Z, Hoatson G (2002) Magn Res Chem 40:70
31. Strandberg E, Wadhwani P, Tremouilhac P, Dürr UHN, Ulrich AS (2006) Biophys J 90:1676
32. Afonin S (2004) Structural studies on membrane-active peptides in lipid bilayers by solid state ^{19}F-NMR. PhD thesis. Institute of Molecular Biology. Friedrich-Schiller-Universität, Jena, p 98
33. Dürr UHN (2005) Solid-state ^{19}F-NMR studies on fluorine-labeled model compounds and biomolecules. PhD thesis. Institute of Organic Chemistry. Universität Karlsruhe (TH), Karlsruhe, p 163
34. Steiner H, Andreu D, Merrifield RB (1988) Biochim Biophys Acta 939:260
35. Blazyk J, Wiegand R, Klein J, Hammer J, Epand RM, Epand RF, Maloy WL, Kari UP (2001) J Biol Chem 276:27899

36. Hallock KJ, Lee DK, Ramamoorthy A (2003) Biophys J 84:3052
37. Mani R, Waring AJ, Lehrer RI, Hong M (2005) Biochim Biophys Acta 1716:11
38. Mani R, Buffy JJ, Waring AJ, Lehrer RI, Hong M (2004) Biochemistry 43:13839
39. Grage SL, Wang J, Cross TA, Ulrich AS (2002) Biophys J 83:3336
40. Möllhoff M, Sternberg U (2001) J Mol Model 7:90
41. Yamaguchi S, Huster D, Waring A, Lehrer RI, Kearney W, Tack BF, Hong M (2001) Biophys J 81:2203
42. Wimley WC (2003) Curr Opin Struct Biol 13:404
43. Montoya M, Gouaux E (2003) Biochim Biophys Acta 1609:19
44. Grotenbreg GM, Timmer MS, Llamas-Saiz AL, Verdoes M, van der Marel GA, van Raaij MJ, Overkleeft HS, Overhand M (2004) J Am Chem Soc 126:3444
45. Maxfield FR (2002) Curr Opin Cell Biol 14:483
46. McInnes C, Kondejewski LH, Hodges RS, Sykes BD (2000) J Biol Chem 275:14287

Top Curr Chem (2008) 273: 155–181
DOI 10.1007/128_2007_17

Published online: 23 April 2008

Modulation of the Bioactive Conformation of Transforming Growth Factor β: Possible Implications of Cation Binding for Biological Function

Eduard V. Bocharov[1] (✉) · Konstantin V. Pavlov[2] · Marcel J. J. Blommers[3] · Tudor Arvinte[4] · Alexander S. Arseniev[1]

[1]Shemyakin-Ovchinnikov Institute of Bioorganic Chemistry, Russian Academy of Science, ul. Miklukho-Maklaya, 16/10, 117997 Moscow, Russia
bon@nmr.ru

[2]Frumkin Institute of Electrochemistry, Russian Academy of Science, 117071 Moscow, Russia

[3]Novartis Pharma AG, 4002 Basel, Switzerland

[4]School of Pharmacy, Department of Pharmaceutics and Biopharmaceutics, University of Geneva, 1211 Geneva, Switzerland

Abstract In any organism, very precisely adjusted interaction and exchange of information between cells is continuously required. These cooperative interactions involve numerous cytokines, acting through corresponding sets of cell-surface receptors. The transforming growth factor β (TGF-β) superfamily includes a variety of structurally related multifunctional cytokines that play critical roles in maintaining cellular homeostasis and controlling cell fate. Response of a cell to a specific signal it receives should depend upon the current state of the environment, including concentrations of biologically relevant ions. One of the most biologically active ions, calcium, acts upon a specific calcium signaling system that operates over a wide temporal range and regulates many cellular processes in continuous "cross-talk" with the TGF-β signaling system. In addition to that, the structural and dynamical properties of TGF-β molecules, along with detected direct interaction of them with the biologically relevant cations suggest another level of fine regulation of TGF-β activity. The fact that both mono- and divalent cations bind in the same low-affinity sites implies that some competition of cations for interaction with TGF-β can also occur in vivo, contributing to the diversity of TGF-β biological functions.

Keywords Cation · Cell-signaling · Dynamics · Structure · Transforming growth factor

Abbreviations

BMP	Bone morphogenetic protein
ecActR2B	Ectodomain of type IIB receptor of activin
ecBR1A	Ectodomain of type IA receptors of BMP-2
ectodomain	Extracellular domain of a receptor
ecTβR1	Ectodomain of type I receptor of TGF-β isoforms
ecTβR2	Ectodomain of type II receptor of TGF-β isoforms
HSQC	Heteronuclear single-quantum coherence NMR experiment
TβRI	Type I receptor of TGF-β ligand
TβRII	Type II receptor of TGF-β ligand
TGF-β	Transforming growth factor β

1 Introduction

The transforming growth factor-β (TGF-β) pathway occupies a central position in the cell-signaling networks [1, 2]. Alterations in the TGF-β pathway in human have been implicated in numerous disease states, including immune suppression, cancer, osteoporosis, hypertension, atherosclerosis, and fibrotic disease of the kidney, liver, and lung [3–11]. TGF-β ligands are multifunctional cytokines that share a common structural scaffold and take part in controlling various cellular responses, including proliferation, differentiation, growth, motility, adhesion and apoptosis in a tissue-specific manner [1–4]. The TGF-β superfamily includes TGF-β isoforms, bone morphogenetic proteins (BMP), growth and differentiation factors, activins/inhibins and some other structurally and functionally related proteins. They all form ~25 kDa disulfide-linked dimers, usually homodimers, though some of them are capable of forming active heterodimers [12, 13]. In terms of monomer fold, the entire family shares identical disulfide connectivity associated with a unique cysteine-knot structure formed by six conserved cysteine residues and similar overall topology including two pairs of antiparallel β-strands and one major α-helical unit. The two monomers in the TGF-β dimer interact through a relatively small interface that was shown to be highly mobile for some representatives of the superfamily [14–17]. The monomer fold of the ligands is often described as an outstretched "hand" with major "*wrist*" and "*knuckle/fingertips*" receptor-binding epitopes (Fig. 1) [12, 18]. Each of the two *knuckle* epitopes in the ligand dimer is located on an individual subunit, whereas the surface of each of the two *wrist* epitopes is shared between the two subunits. For propagating signals across the cell membrane, the members of the TGF-β superfamily require at least two structurally related types of receptors, type I (TβRI) and type II (TβRII), both having a short extracellular domain (ectodomain), a single membrane-spanning region and an intracellular serine/threonine kinase domain [12, 18, 19]. In the ligand-free state, the receptors of both types form homodimers on the cellular membranes [19]. The sequence of events associated with the receptor binding varies somewhat between different representatives of the family, however, the most typical sequence, which is characteristic for TGF-β isoforms, is as follows [12, 19]. During the signaling process the TGF-β ligand binds first to its type II receptors through the *knuckle/fingertips* epitope and then to type I receptors through the *wrist* epitope, forming ligand-mediated TβRI-TβRII heterodimers. In the heteromeric complex of TGF-β with the receptors, the kinase domain of TβRII phosphorylates TβRI, which in turn phosphorylates downstream intracellular signaling components. There are additional membrane proteins that can participate in formation of the signaling complex and regulation of signal transduction on both the extracellular and cytosolic side [20, 21]. This includes betaglycan, also known as the type III receptor. However, it was only demonstrated to be necessary for recognition of TGF-β2, and its main role presumably is to correctly present TGF-β ligands to the type II receptor [19].

The biological response to the TGF-β signaling in different tissues is modulated by a large number of cofactors, ranging from proteins to inorganic ions, both intra- and extracellular [2, 9, 11, 12]. One of the important cofactors is calcium that controls a variety of cell functions in an organism including secretion, growth, differentiation and mobility [22–24]. It was shown that during tissue development Ca^{2+} transients generated by both intracellular and extracellular Ca^{2+} stores (acting either separately or in combination) have multiple roles in cell differentiation, including modulation of TGF-β signaling [25, 26]. TGF-β ligands, in turn, are involved in modulating intracellular calcium concentration; they have been reported both to increase intracellular calcium [27, 28] as well as to inhibit calcium mobilization stimulated by other growth factors [29, 30]. On the extracellular side, Ca^{2+} ions are known to affect TGF-β signaling in various cells [31–34]. Moreover, chelation of extracellular Ca^{2+} was shown to fully inhibit the cell response to TGF-β ligands in several cases [35–37]. Regulation of the cell response to TGF-β by calcium might be especially important in the case of bone tissues, where the local calcium concentrations at the membrane surface are known to reach tens of millimoles, which is one order of magnitude above normal physiological levels (e.g., extracellular Ca^{2+} concentration underneath a resorbing osteoclast can reach 40 mM) [11, 38, 39]. Elevated local extracellular Ca^{2+} concentrations were also shown to exist in blood near the sites where calcium is translocated across epithelial interfaces, or within inflammatory fluids due to release of calcium from intracellular stores during setting of inflammation and apoptosis [23]. TGF-β superfamily members also play a role in these processes [1–4]. In pharmaceutical compositions, the joint application of TGF-β and different calcium salts synergistically stimulates the bone and tissue repair [40–43]. Additionally, it has been demonstrated that certain pharmaceutical compositions containing TGF-β ligands, especially TGF-β3, and calcium salts are very stable chemically, whereas in the absence of Ca^{2+} ions chemical degradation is substantial [43]. The body of these facts suggests a possibility of direct interaction of Ca^{2+} ions with TGF-β, and raises the issue of specific binding of the cations. However, binding of other biologically relevant ions also cannot be ruled out a priori.

At the physiological level, regulation by calcium appears to be interconnected with the sodium concentration in many tissues [23, 44]. For example, in the kidney, sodium reabsorption helps to set the transepithelial potential and the passive paracellular movement of calcium. For physicians and scientists interested in kidney disease, TGF-β representatives are perhaps the most important growth factors yet identified [9, 10]. Variations of sodium intake are believed to affect the balance between TGF-β and nitric oxide synthase, thus creating a negative feedback loop, because TGF-β increases expression of the synthase and nitric oxide inhibits production of TGF-β. Changes in this feedback system have been observed in salt-sensitive hypertension and appear to impact the end organ damage, particularly the kidney [8, 9]. Similarly, the expression of TGF-β was shown to be elevated in the damaged ventricles of the heart after an infarct in comparison with intact ventricles; and the time course of the release of TGF-β isoforms was shown to be different for each isoform [7]. It was shown that dietary salt increased production of TGF-β independently of blood pressure. At the same time, TGF-β may control blood pressure levels by various mechanisms [45]. Although there are many mechanisms of regulation of the cell functions by ionic strength, which is mostly controlled by sodium concentration in the extracellular environment, the effects of direct interaction of Na^{+} ions with TGF-β ligands also cannot be ignored.

As described below in detail, direct interaction of TGF-beta molecules with metallic cations was indeed observed. Moreover, the interaction occurs in the biologically relevant ranges of concentrations, and the cation binding sites are located in the biologically sensitive parts of the molecule. There are indications that TGF-β ligands experience large-scale conformational transitions during signal transduction. In order to assess the effects of cation binding on TGF-β activity, possible conformational rearrangements experienced by TGF-β ligands and their positioning with respect to receptors during signal transduction need to be understood. This can be achieved by comparison of the available structural, dynamics and kinetics data on interaction of different TGF-β representatives with the receptors.

2
Molecular Rearrangements of TGF-β During Signaling Complex Formation

2.1
Principal Conformational States, Observed for TGF-β Representatives

Over the last decade, a substantial amount of information about properties and spatial structures of different representatives of the TGF-β superfamily, both free and in complexes with ectodomains of their transmembrane receptors, have been accumulated. Consequently, a number of important steps towards understanding initiation and actuation of TGF-β signaling, as well as the biological response to the signal transduction and interplay with other signaling pathways have been made [2, 12, 18, 46–50].

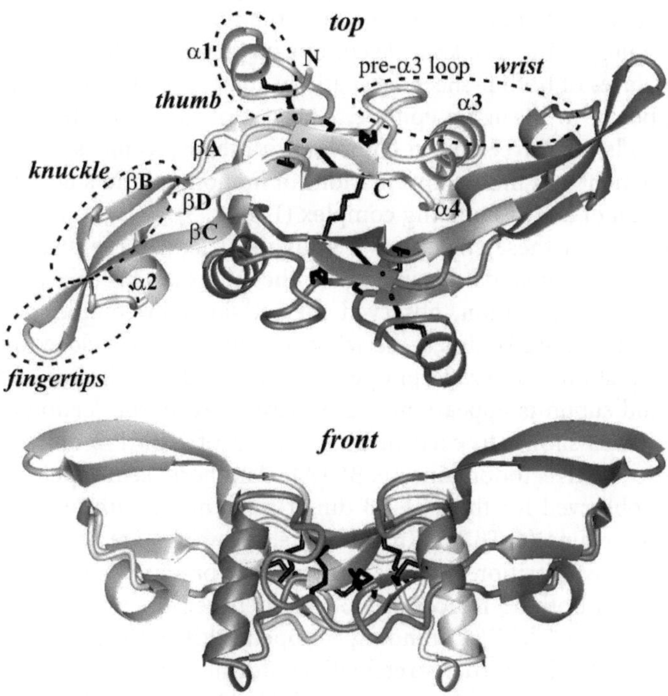

Fig. 1 Ribbon presentations of spatial structure of the TGF-β3 dimer. Secondary structure elements and chain termini for the one monomer are labelled on the *top* view. TGF-β type I receptor binding epitope, designated as "*wrist*", TGF-β type II receptor binding epitopes, named as "*knuckle/fingertips*", and an additional "*thumb*" epitope are circumscribed (on one monomer) by *dashed ovals*. The S–S bonds are shown in *black*

The available NMR and X-ray data provide important insight concerning the nature of conformational transitions that can be experienced by the TGF-β dimer and their significance for the biological function of the ligands. Structurally, they consist of a cysteine-knot core of the dimer and two wing-like extensions consisting of β-sheets stretched outwards. The wings are not planar, having distinctly concave and convex surfaces. The spatial structure, typical for TGF-β isoforms, is shown in Fig. 1. Each monomer of the TGF-β dimer adopts an extended fold made up of four helices and four β-strands arranged in two irregular antiparallel β-sheets, βA/βB and βC/βD [51]. The long $\alpha 3$ helix (the α-helical unit found in all representatives of the family) from one monomer can pack against the other monomer in its concave surface. The *wrist* epitope for TβRI comprises the residues of $\alpha 3$ helix and its pre-helix loop of one subunit, and a portion of the concave surface of the wing of the other subunit, located in the vicinity of the *fingertips* region [52]. According to [52–54], there are "hot spots" for interaction with TβRI both in the pre-$\alpha 3$ helix loop and on the wing surface (hydrophobic cavity near $\alpha 2$ helix). The interaction with TβRII in different species of the TGF-β family was shown to occur through adjacent *knuckle* or *fingertips* epitopes [15–17, 55]. It was determined that the residues of the loops at the tips of both β-sheets form the *fingertips* epitope [16]. The *knuckle* epitope mapped [15, 17, 55] onto the outer convex surfaces of both β-sheets is spatially adjacent to the *fingertips* epitope and is opposite to the *wrist* epitope. Analysis of the available structures of TGF-β ligands, both free and in complex with their receptors, reveals different arrangement and mutual orientations of monomers and suggests putative conformation of active signaling complex (Figs. 2, 3).

Comparison of these structures highlights at least three distinct conformations of TGF-β ligands depending on the environment (Figs. 2, 6a,b). In a "compact" conformation, observed for activin A in complex with TβRII ectodomain (ecActR2B) dimer, the wings of the ligand are close together. In this conformation, the *wrist* epitope for TβRI made by the residues of different ligand subunits appears not to be assembled in the "canonical" form, in which it was shown to exist in the complex of a BMP-2 dimer with two monomeric TβRI ectodomains (ecBR1A). Another, "semi-open" conformation, was observed for the TGF-β3 dimer in complex with two monomeric TβRII ectodomains (ecTβR2). In this state, the wings are somewhat drawn apart, the relative position and orientation of the pre- and post-$\alpha 3$ helix loops suggesting that both $\alpha 3$ helices are located between the wings and are adjacent to each other. The TβRI binding epitope in this conformation is still not fully assembled. The third, "stretched" conformation, which was observed, for example, for crystal structures of free TGF-β2 and TGF-β3 dimers and for the BMP-2 dimer with two monomeric TβRI ectodomains, is fully stretched, with the two subunits of the ligand nearly parallel to each other. In this case, all parts of the *wrist* epitope are properly arranged to bind the TβRI receptor. Besides the angle between the long axes of the wings, which cross in the

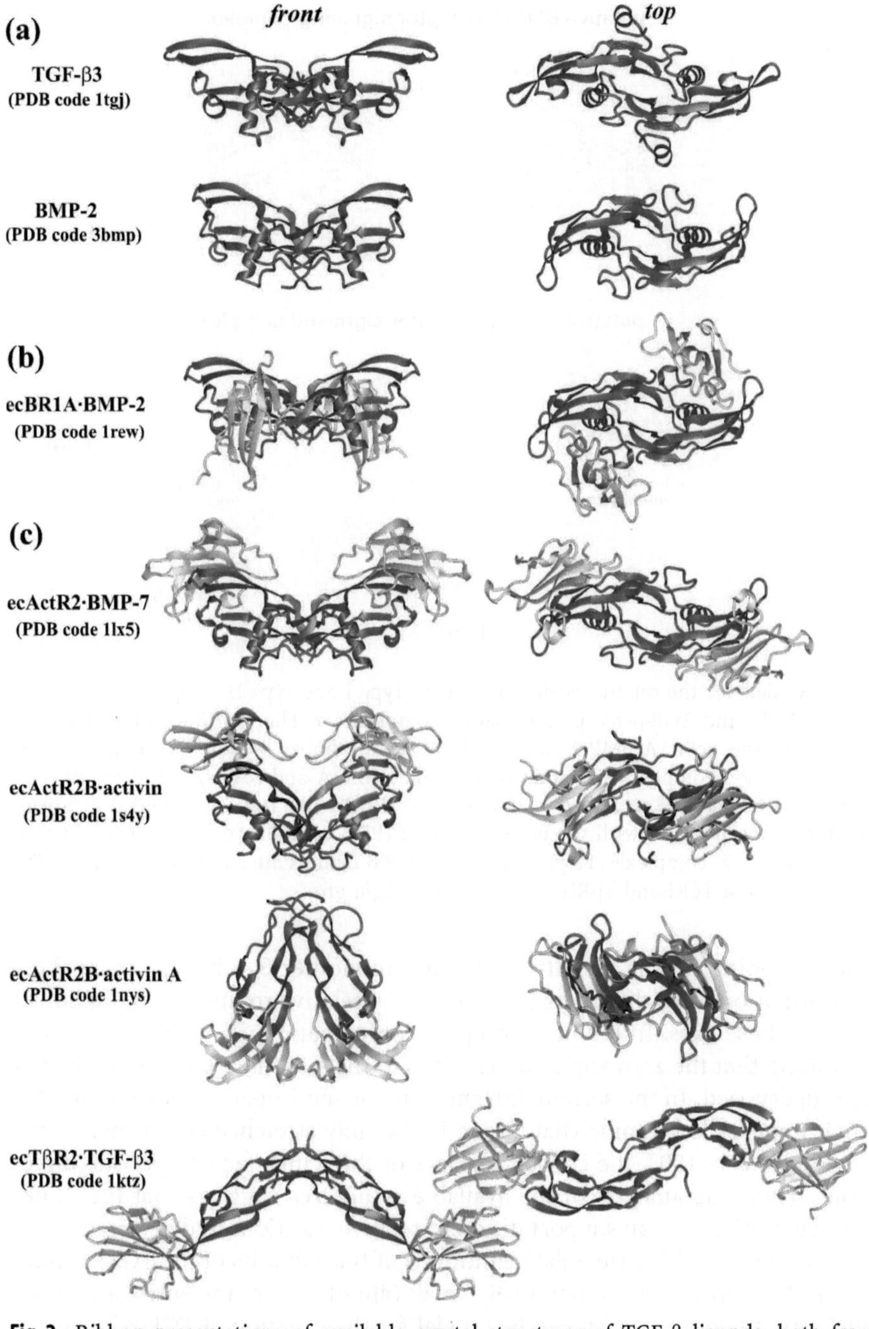

Fig. 2 Ribbon presentations of available crystal structures of TGF-β ligands, both free (**a**) and in complexes with their TβRI (**b**) and TβRII (**c**) receptors. The *front* and *top* views are shown. The TGF-β ligands are shown in *dark gray*. The ectodomains of TβRI and TβRII receptors are highlighted in *light gray*

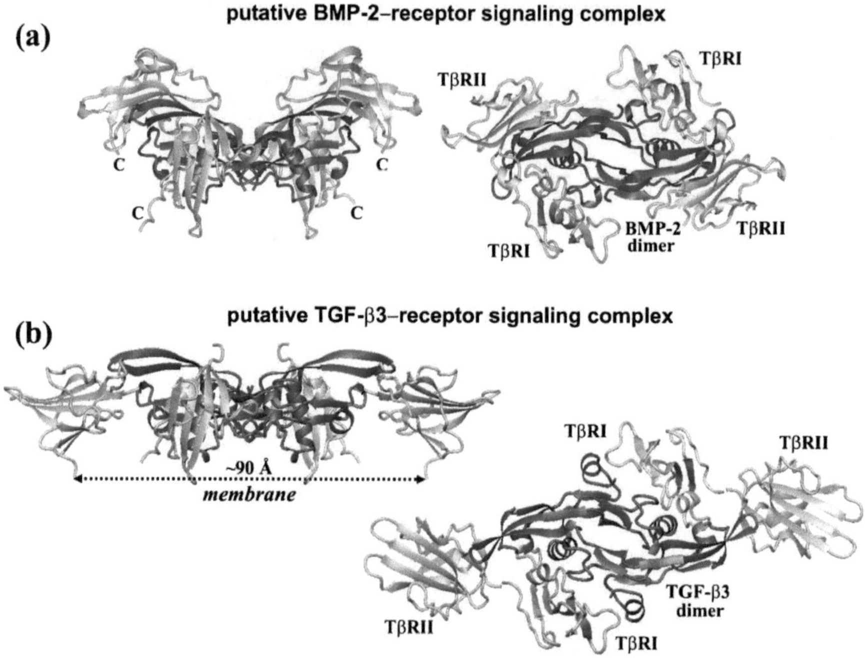

Fig. 3 Models for the relative positioning of the type I and type II receptor ectodomains in the BMP- and TGF-β-receptor signaling complexes. **a** The ActR2 ectodomains were docked to the ecBR1A·BMP-2 complex (PDB code 3bmp) by superposition with the ecActR2·BMP-7 complex (PDB code 1lx5). **b** The ecBR1A and ecTβR2 ectodomains were docked to the crystal structure of free TGF-β3 (PDB code 1tgj) in extended conformation by superposition with the ecBR1A·BMP-2 (PDB code 3bmp) and ecTβR2·TGF-β3 (PDB code 3ktz) complexes, respectively. The TGF-β ligands are shown in *dark gray*. The ectodomains of TβRI and TβRII receptors are in *light gray*

"pivotal point" in the vicinity of the intermonomer S–S bridge, these three conformations also differ from each other by relative rotation of the subunits around these axes. In the first, compact conformation, the angle is minimal, assuming that the zero angle corresponds to the situation when two subunits are superposed. In the second, intermediate or semi-open conformation, this angle appears to be somewhat larger. In the fully stretched conformation the angle is nearly 180°, the convex surfaces of the wings facing opposite directions. Thus, the analysis of the available structures suggests that the dimer interface of TGF-β can support different relative positions and orientations of the subunits. In this case, relative motions of the subunits of the TGF-β dimer connected through a comparatively small (about 10% of the entire accessible surface) but complex joint can be crucial for the function of TGF-β ligands.

This is supported by our data obtained for TGF-β3 dynamics in solution [14], according to which the average conformation of the TGF-β3 dimer is less extended than its conformation in crystal [51], as evidenced by the ratios of the principal components of the rotational diffusion tensor D_x/D_z and D_y/D_z derived from the relaxation data assessed at about 0.75 compared to approximately 0.44, estimated from the TGF-β3 crystal structure of the free TGF-β3 dimer. The dimeric interface region of TGF-β3 was shown to be very flexible, both on the fast and slow time scales (Fig. 6c,d). Similarly, slow exchange was observed in the dimer interface for TGF-β1 in solution, namely within pre- and post-α3 helix loops [56]. The dynamics study also implies that certain properties of the TGF-β3 dimer should be very sensitive to mild variations of local environment, which might be coupled with binding the receptors [14, 57]. Moreover, the pattern of internal dynamics in the structure identifies potential "hot spots" of binding free energy and reveals the importance of conformational entropy in the interaction of TGF-β3 with the receptors. In particular, the α3 helices of the dimer, involved in forming the *wrist* epitopes, can undergo a helix-coil transition in aqueous solution, whereas in a nonpolar environment they prefer to fold into long α-helices, which interact with the TβRI ectodomains [14].

In summary, the three principal conformations, adopted by different TGF-β representatives, that can be described as "compact", "semi-open", and "stretched" (Fig. 6a,b), differ in mutual orientation of monomers and in the degree to which the receptor binding epitopes, in particular the *wrist* epitopes, are assembled. A possible interpretation is that they reflect different architectures of active signaling complex for individual representatives of the family, determining diversity of biological response [58] (many cases of cross-interaction between different TGF-β ligands and their receptors have been reported [6, 12, 19]). However, structural homology between TGF-β ligands, mobility of their dimeric interface, and observed transient nature of the TβRI epitope lead to a fairly simple assumption that all the three principle conformations are inherent to the entire TGF-β family. Transitions between these conformational states, experienced by representatives of the family, can be triggered by receptor recognition events. Similar assumptions were recently used to explain the cooperative interaction of the TGF-β ligands with the receptors [15]. In other words, the different conformations observed for the dimers of activin, BMP and TGF-β isoforms correspond to different "frames in the movie" showing molecular interactions of TGF-β ligands with receptors, rather than being specific to individual representatives of the family. Thus, each of the TGF-β crystal structures corresponds to a certain energy minimum of the ligand, "frozen" in one of the conformational states, out of the entire spectrum of conformations inherent to the protein.

2.2 Potential Scenarios of the Initial Stages of TGF-β Signaling Complex Formation

The observed conformational states and dynamical properties of TGF-β ligands suggest a dynamical model of their cooperative interaction with receptors, when binding of one receptor initiates folding of the epitope for the other. Principally, the ligand can approach the receptors from two sides—from the bulk solution and from the side of the membrane. These two cases differ by 180° rotation of the ligand around the axis parallel to the membrane surface. However, the two scenarios are not fundamentally different and explain the majority of the available facts equally well. A hypothesis of membrane-restricted ligand immobilization recently outlined in [15] considers the initial receptor recognition by the ligand in the compact conformation occurring with the pivotal point in the immediate vicinity of the membrane surface. However, it was found with the use of plasmon resonance spectroscopy that the dimer-to-dimer interaction of TGF-β isoforms with artificially dimerized ectodomains of the TβRII receptor is about equally effective in both cases—when the ligand is immobilized on the surface and the receptor dimer (but not monomer) is in the bulk solution and vice versa [59–61]. Moreover, in both cases the kinetics of the dimer-to-dimer (but not of dimer-to-monomer) interaction includes a rearrangement step, which can be a conformational transition of the receptor-ligand complex involving transfer of the ligand into an intermediate state after initial recognition [59–61]. Hence, the membrane-restricted setting of the ligand is not necessary for TβRII recognition by TGF-β, but the conformational transitions during recognition are essential. Therefore, the possibility of formation of the TGF-β signaling complex with the ligand approaching the receptors directly from the bulk solution appears more likely. Below we describe the sequence of events corresponding to this case.

Initially, the ligand recognizes the TβRII dimer by its accessible *knuckle/fingertips* epitopes and assumes the compact conformation similar to the one observed in the activin/ActRII complex (Fig. 4a). Binding of the TGF-β dimer to the receptor homodimer causes a conformation transition, during which the receptor dimer dissociates and the resulting monomers diverge somewhat. Accordingly, the wings of the ligand move apart. At the same stage, the wings are rotated in the opposite directions around the axes, perpendicular to the joint. When the mutual orientation of the monomers and the angle between the wings become adequate, the epitopes for the type I receptors start forming and become accessible (Fig. 4b). Moreover, since the *knuckle/fingertips* ligand epitopes for TβRII are bound to the receptors, mov-

Fig. 4 Hypothetical dynamical model for interaction of TGF-β ligands with receptor dimers and formation of the signaling complex on the cell surface. Formation of the active TGF-β signaling complex after ligand binding to TβRII is decomposed into three stages, corresponding to compact (**a**), semi-open (**b**), and extended (**c**) conformations of the ligand. The axes, around which the monomers of the TGF-β dimer rotate, are shown on (**b**). *Pairs of arrows* indicate the direction of motion of the ligand monomers with respect to each other (divergence of the axes and rotation around them); the *single larger arrow* shows tumbling of the dimer as a whole during ultimate formation of the active signaling complex ▶

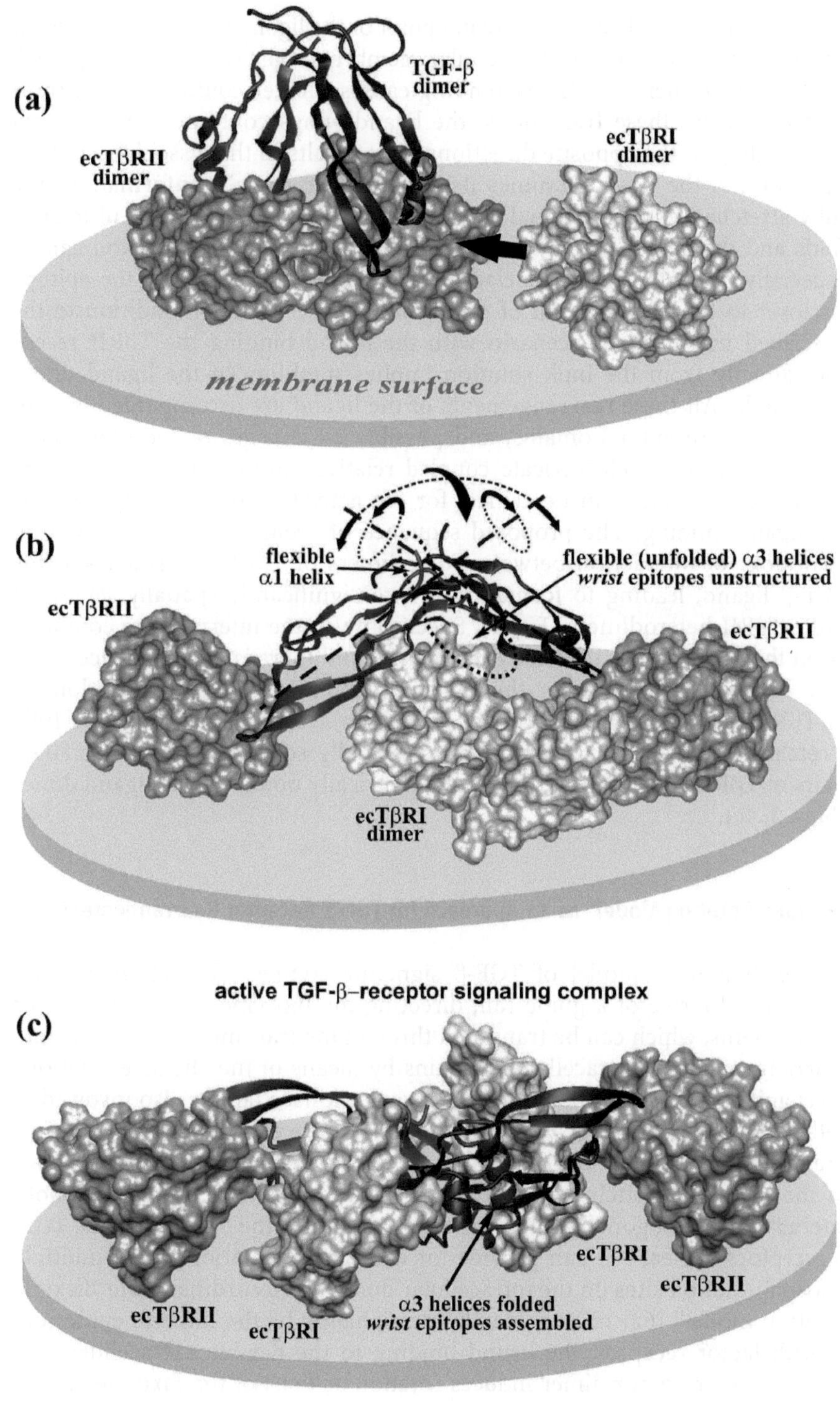
(a)
TGF-β dimer
ecTβRII dimer
ecTβRI dimer
membrane surface
(b)
flexible α1 helix
flexible (unfolded) α3 helices
wrist epitopes unstructured
ecTβRII
ecTβRII
ecTβRI dimer
active TGF-β–receptor signaling complex
(c)
ecTβRI
ecTβRII
ecTβRII
ecTβRI
α3 helices folded
wrist epitopes assembled

ing the wings apart brings the pivotal point of the ligand and pre-folded *wrist* epitopes for TβRI closer to the cellular membrane. The subsequent approach of the TβRI dimer and its pre-binding causes further conformational transitions. During these transitions, the ligand wings continue moving apart and rotating in the opposite directions. This results in the dissociation of the TβRI dimer; the ligand assumes its final "spread-eagle" conformation (the fully stretched conformational state), in which the *wrist* epitope ultimately folds and the active signaling complex forms (Fig. 4c). The rotation can be especially important for dissociation of the TβRI dimer, where the epitope is closer to the pivotal point of the ligand (to the joint). In addition to the described motions, the scenario with the ligand binding the TβRII receptors directly from the bulk solution implies tumbling of the ligand dimer as a whole. All these rearrangements of the ligand are accompanied by motions of the receptor domains, and possibly by changes of receptor-ligand contact surfaces. Such intricate coupled relative motions of the monomers in the TGF-β dimer can be critical for the activation of the receptors during ligand binding. The proposed sequence of events explains the existing "paradox" of interaction between homodimers of TβRI and TβRII and the TGF-β ligand, leading to formation of two significantly spatially separated TβRI-TβRII heterodimers (Fig. 3). Specifically, in the intermediate conformation the TβRII dimer is already dissociated and there is enough space available between the wings for the smaller (compared to TβRII) ectodomain of TβRI to access both *wrist* epitopes (Fig. 4b). After transition to the fully stretched conformation, there are two spatially separated (by about 70 Å) pairs of TβRI-TβRII heterodimers, symmetrically bound to the ligand dimer (Figs. 3, 4c).

2.3 "Flexible Rotation Model" as an Approach for TGF-β Receptor Rearrangements

In the dynamical model of TGF-β signaling complex formation, the ligand plays the role of a guide rail, directing the movements of the receptor ectodomains, which can be translated through the transmembrane parts into rearrangements of intracellular domains by means of the "flexible rotation" mechanism recently proposed for the tyrosine kinase family, also involved in cell-signaling processes [62]. These motions of the extracellular, transmembrane and intracellular domains of the receptors may include rotation, change of the inclination/pitch, and shifts relative to the cellular membrane, both laterally and orthogonally. The implication of this for the function of the TGF-β receptors is clear—it can provide for correct presentation of the multiple phosphorylation sites on the intracellular domains. According to the "flexible rotation model" [62] of transmembrane signaling by the dimeric epidermal growth factor receptor, the ligand binding to the flexible extracellular domains of the receptor dimer induces rotation or twist of the juxtamembrane regions. This results in the rotation and twist of the helical transmembrane domains and dissociation of the dimeric, inactive form of the intracellular domains, causing change in the accessibility of the intrinsic tyrosine kinase to multiple tyrosine residues. The authors also suggested that such flexible rotations might be a common property of many cell-surface receptors. Indeed, comparisons of the structures of the free and ligand-bound aspartate receptors (dimeric transmembrane signaling protein) identified that relatively small conformational changes of the extracellular domain induce considerable rotation between the subunits of the receptor dimer [63]. The likelihood of the hypothesis about the universal role of rotational coupling of the transmembrane and kinase domains increases in the light of the accumulating findings of similar mechanisms in various signaling systems [64, 65]. The case of serine/threonine kinase receptors for TGF-β family ligands appears to be more complex, because two types of homodimeric receptors normally interact with the homodimeric ligand to form two spatially separated heterodimers. For this reason, a wider range of spatial rearrangements may be required for functioning of this complex articulation system. However, these rearrangements can be decomposed into distinct patterns of sequential motions similar to the ones proposed for other types of transmembrane receptors.

In particular, the rotation and swinging of the extracellular domains coupled with rotation and tilting of the transmembrane domain might occur during dissolution of the receptor homodimers (as is very schematically illustrated by Fig. 5) and formation of heterodimers and can be important at the initial and final stages of interaction of the TGF-β dimer with both receptor dimers. Such motions and conformational transitions of the extracellular and transmembrane domains cannot occur without similar motions of the intracellular domains, resulting in changes of their mutual orientation, the points of contact between them, and hence of the distance between the

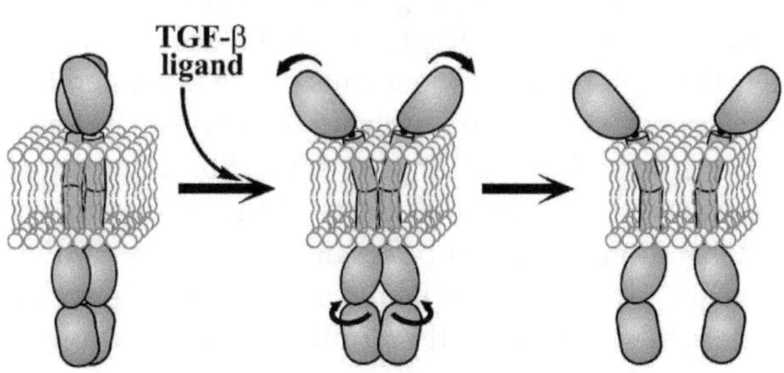

Fig. 5 Possible motions of the type II receptors during interaction with the TGF-β ligand. Binding of the ligand to the TβRII receptor homodimer induces rotation and swinging of the extracellular domains, which are translated through the transmembrane parts into rearrangements of intracellular domains by means of a "flexible rotation" mechanism, finally causing dissolution of the dimer and spatial separation of the receptor monomers

juxtamembrane regions. Another important pattern can include coupling of the orthogonal shift of the transmembrane region with respect to the membrane with the rotation of the intracellular domains. According to putative structures of BMP-2– and TGF-β3–receptor signaling complexes and the conformation sequence shown on the Figs. 3, 4, when both TβRI and TβRII are bound to the ligand, the opening of the ligand wings can result in either pulling the TβRII out of the membrane, or plunging TβRI into the membrane. Small as these shifts might be, they can cause significant rotation of the intracellular domains, for example, change in the length of a transmembrane helix by one residue induces rotation of the receptor intracellular domains by approx. 100°. Indeed, progressive deletions of defined numbers of residues in the helical transmembrane domains of mutant TGF-β type II receptors rendered them constitutively active [66, 67]. It has been demonstrated that for TGF-β the relative rotation of the receptor intracellular domains by approx. 100° is required for signal transduction. Similarly, periodic activation of the Neu receptor induced by transmembrane-shift mutations has been observed in [64]. The authors also quote a number of examples where similar effects were observed for different signaling receptors. Another pattern, which might be specific for TGF-β and occur during the intermediate stage of interaction with the receptors, involves spatial separation of the TβRII monomers (up to ~90 Å for TGF-β isoforms, Fig. 3b). At this stage, relative rotation of the ligand subunits in the opposite directions around their long axes as the ligand assumes its final extended conformation is also possible (to account for the difference in the monomer orientations between the intermediate and final conformations) (Fig. 4b). Moreover, this might result in coupled rotation of the extracellular domains of both TβRII monomers, which can facilitate the diffusion or dragging of the TβRI dimer towards its epitope. The possibility of interaction between TβRI and TβRII in the absence of ligands is revealed by the existence of inactive TβRI-TβRII oligomers [19], which are activated by a reorientational rotation triggered by ligand binding [66, 67].

Finally, there is a distinct coupling between the conformational changes and movements of extracellular receptor domains and those of the intracellular kinase parts through the rearrangements of the transmembrane segment. Variations of the parameters, characterizing orientation and position of each part (angles, translational shifts) impose some limitation on the allowable values of the parameters of the other parts. If the assumption about the sequence of conformational changes being inherent to the TGF-β superfamily is correct, the major function of the ligand at the early stages of cell-signaling is that of a guide rail, imposing further limitations on possible movements of the receptor parts and thus defining the scenario of the transmembrane signaling. When these limitations are compatible with the intrinsic limitations of the receptors, some response to the ligand binding is possible. The particular response observed in each individual case might be different, depending upon the bound ligand and the environment.

3
Binding of Cations by TGF-β3: Possible Implications for Biological Function

3.1
TGF-β3 Cation-Binding Sites

In our studies of the TGF-β3 dimer in solution, we detected the direct interaction of several metallic cations, such as Ca^{2+}, Na^{+}, and Mn^{2+} with the aid of NMR spectroscopy. In particular, chemical shift variations in response to changes of Na^{+} and Ca^{2+} ion concentrations and line broadening due to the interaction with paramagnetic Mn^{2+} ions were observed in the heteronuclear two-dimensional ^{1}H-^{15}N HSQC spectra and mapped onto the spatial structure of the TGF-β3 using color-coded ribbon and surface representations (Figs. 6e, 7a,b).

A clear pattern, indicative of four distinct localized cation-binding sites per monomer can be seen on the diagrams. The "*N-terminal*" site is formed by the residues of the N-terminus and βC strand, including the residues Asp 3, Asp 82 and Glu 84, the carboxylate side chains of which have a high ability to coordinate cations. Within the "*H1-loop*" site following the $\alpha 1$ helix there are glutamate residues Glu 12 and Glu 13 as potential cation chelators. The "*H2*" site is located in a cavity near the $\alpha 2$ helix, and formed by the side chains of the Asp 23, Asp 27 and Tyr 21 residues, which are highly conserved in the TGF-β family. A large area in the interface of the dimer, around the intermonomer S–S bond formed by Cys 77–Cys 77″ residues, is designated as the "*C-terminal*" site. The site includes carboxylate groups of Asp 55 and of the C-terminal Ser 112. Backbone and side chain polar groups of some other residues in the vicinity of the identified cation-binding sites may also play a role in interaction with cations. Putative locations of all the cation-binding sites based on our experimental data are within the areas having negative electrostatic potentials calculated for the physiological pH (Fig. 7d). The cations can be placed within the binding sites (Fig. 7d) in such a way that the distances from the center of the ion to the carboxylate oxygen atoms would be consistent with the radii of the cation coordination spheres (2–3 Å typical for physiologically relevant cations [22]).

According to the available structures of BMP, two out of the four cation-binding sites identified for TGF-β—*C-terminal* and *H2* ones—are also present in BMP molecules. Besides these two sites, analysis of the crystal structures of BMP-2 (PDB code 3bmp) and BMP-7 (PDB code 1lxi) suggests possible existence of an additional divalent cation-binding site (per monomer) in the *fingertips* region near their *wrist* epitope, consisting of the side chains of Asp 93, Glu 94 and Asp 118, Asp 119, respectively. Similarly, the spectrum of cation binding sites can vary between representatives of the TGF-β family reflecting their specific biological function.

Ca^{2+} and Na^{+} dissociation constants were assessed from the chemical shift titration curves. The *H1-loop* site appears to have the lowest affinity (about 50 and 80 mM, respectively,) for the cations, in agreement with the small values of changes of the signal intensity caused by Mn^{2+} addition. The Ca^{2+} and Na^{+} binding constants for the other three sites are about 15 and 20 mM, respectively, which is in the physiologically relevant range for certain tissues, therefore these sites can be described as "low-affinity ion traps". It should

Fig. 6 TGF-β conformational flexibility and cation binding to the TGF-β dimer. **a** Principal conformational states observed for TGF-β representatives. The "compact", "semi-open", and "stretched" conformations of the TGF-β ligand are presented using crystal structures of ecActR2B·activin A complex (PDB code 1nys), ecTβR2·TGF-β3 complex (PDB code 1ktz) and free TGF-β3 (PDB code 1tgj) dimer, respectively. The monomers in TGF-β dimers are shown in *magenta* and *blue*. **b** Comparison of TGF-β principal conformations by superposition of the monomers shown in *magenta*. Positions of the other monomers of TGF-β dimers in compact, semi-open, and stretched conformations are depicted in *green*, *cyan*, and *blue*, respectively. The conformations differ in mutual orientation of monomers in the ligand dimer and in the degree to which the *wrist* receptor binding epitopes are assembled. **c** Slow conformational exchange of TGF-β3 in solution mapped to its crystal structure (PDB code 1tgj). Residues with micro-millisecond time scale motions identified by exchange contributions R_{ex} to ^{15}N transverse relaxation rates R_2 are colored in *magenta* (for details, see [14]). Residues 6–17 and 48–75 exhibiting strong broadening and pronounced doubling of backbone NH resonances, respectively, indicative of slow conformational exchange are shown in *cyan*. The S–S bonds are highlighted in *orange*. **d** Fast internal dynamics of TGF-β3 in solution mapped to its crystal structure (PDB code 1tgj). Identified cation positions in the TGF-β3 dimer are shown on the ribbon diagram color-coded according to the generalized order parameter S^2 reflecting the fast pico-nanosecond dynamics of the TGF-β3 backbone (for details, see [14]). The cations are presented as *green* balls of 1 Å radius surrounded by *dotted* coordination spheres with the radius of 2.4 Å. The color-scale reflecting variation of S^2 from 0.2 to 0.8 is shown by the *bar*. TGF-β receptor binding epitopes are circumscribed (on one monomer) by *dashed ovals*. The S–S bonds are highlighted in *orange*. **e** Color-coded ribbon representations of TGF-β3 response to cation binding. Generalized chemical shift changes and decrease of signal intensity in response to addition of 10 mM $CaCl_2$ and 1 mM $MnCl_2$, respectively, to the TGF-β3 dimer are mapped to its crystal structure (PDB code 1tgj) using color-coding. The response to cation addition is shown on different monomers of the dimer, $CaCl_2$ on the right one as seen from the *top* view, and $MnCl_2$—on the *left* one. The generalized chemical shift change $\Delta\delta$ was calculated as the module (in Hz) of displacement vector between cross-peak positions in the two-dimensional ^{1}H-^{15}N HSQC spectra acquired for TGF-β3 sample titrated by calcium. The effect of Mn^{2+} was presented as a ratio of cross peak intensities of the amide group in ^{1}H-^{15}N HSQC spectra recorded with ($I_{Mn^{2+}}$) and without (I_0) 1 mM $MnCl_2$. The color-scales are shown on the same bar, reflecting the increase of $\Delta\delta$ from 7 to 40 and decrease of $I_{Mn^{2+}}/I_0$ from 0.9 to 0.4. Secondary structure elements and chain termini for one monomer are labelled on the *top* view. The aspartate and glutamate side chains, as well the C-termini are shown. The side chains forming the identified cation binding sites, as well as the sites themselves, are labelled on one monomer. The *front* view on (**c–e**) is on the *left*, at a scale two-times smaller compared to the *top* view

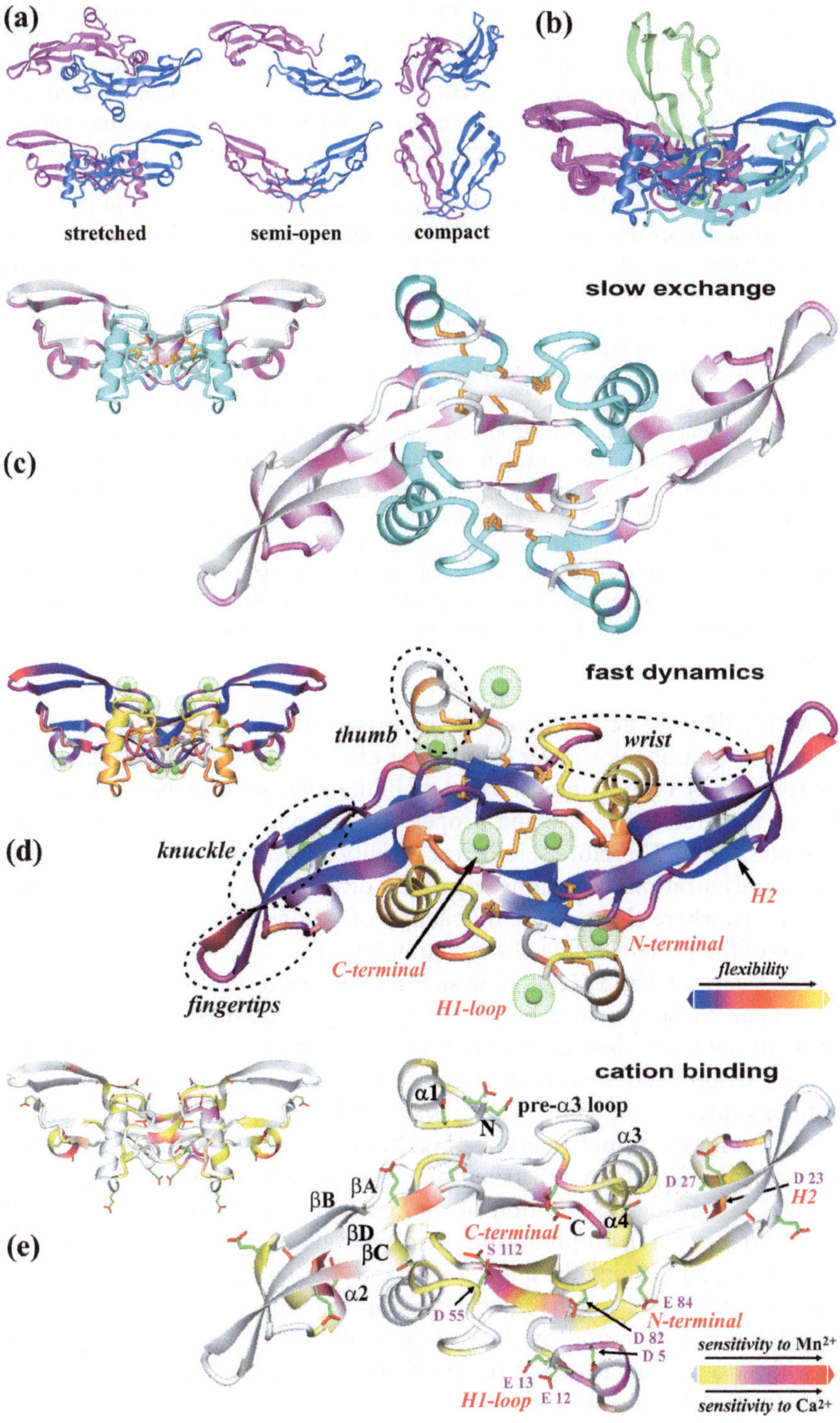
(a)
(b)
stretched
semi-open
compact
slow exchange
(c)
fast dynamics
thumb
wrist
knuckle
(d)
H2
N-terminal
C-terminal
fingertips
H1-loop
flexibility
cation binding
α1
pre-α3 loop
N
α3
βA
βB
D 27
D 23
H2
βD
C-terminal
C
α4
(e)
βC
S 112
α2
D 55
E 84
N-terminal
D 82
D 5
sensitivity to Mn^{2+}
E 13
H1-loop
E 12
sensitivity to Ca^{2+}

Fig. 7 Color-coded surface representations of TGF-β3 response to cation binding and of hydrophobic and electrostatic properties of the molecule. **a** and **b** Generalized chemical shift changes and decrease of signal intensity in response to addition of 10 mM $CaCl_2$ and 1 mM $MnCl_2$, respectively, to the TGF-β3 sample (*front* and *back* views, PDB code 1tgj) are mapped on the surface of one monomer using color-coding. **c** Monomer surface of the TGF-β3 dimer colored according to molecular hydrophobicity potential. The values of molecular hydrophobicity potential are given in conventional units. One dioxane molecule and a number of water molecules (per monomer) observed in the crystal TGF-β3 structure are depicted by balls. **d** Putative cation positions on the monomer (*front* and *back* views) and dimer (*top* and *bottom* views) surfaces of TGF-β3 colored according to electrostatic potential. The cations are presented as *green* balls of 1 Å radius size surrounded by point coordination spheres with radius of 2.4 Å. The values of electrostatic potential are given in units of *e*/Å (*e* is electron charge). The color-coding conventions are shown by bars. On the *front* and *back* views the second monomer is presented by point surface for the best visualization of the dimer interface. The identified cation-binding sites per monomer and per dimer are labelled in red on the *front*/*back* and *top*/*bottom* views, respectively. Possible transitions of cations in the intramolecular hydrophilic tube of the one *C-terminal* site are depicted by the *yellow* arrow in (**d**). The molecular hydrophobicity potential created by protein atoms on the protein solvent accessible surface was calculated as described earlier using homemade software [79]. Spatial distribution of electrostatic potential on the protein solvent accessible surfaces was generated by MOLMOL [80]. The visualizations of surface distributions of molecular hydrophobicity and electrostatic potentials and figures of the molecular structures were obtained with MOLMOL ▶

be noted, that the actual values of the dissociation constants in the physiological conditions are lower than those obtained in the experiment, since the specifics of TGF-β3 imposed certain limitations on the experimental conditions required for measuring proper NMR spectra, in particular, low pH of about 3 [68]. Therefore, biological activity of TGF-β3 will be sensitive to lower concentrations of cations, especially divalent ones, in the physiological conditions, where electrostatic interaction of cations with negatively charged groups of the protein will facilitate cation binding.

The *N-terminal* and *H2* binding sites are formed by small concave pads on the protein surface formed by negatively charged residues (Fig. 7d). The location of the *C-terminal* cation-binding sites in the TGF-β dimer is the most difficult to precisely determine, because the amide groups sensitive to ion additions form two extended surfaces surrounding the intersubunit disulfide bridge within the dimer interface (Fig. 7a,b). A similar pattern of the negative electrostatic potential distribution can be observed in Fig. 7d. According to Fig. 7c, showing distribution of the molecular hydrophobicity potential, these surfaces are hydrophilic and form the walls of two curved hydrophilic tubes (or wells) fringing the intermolecular S–S bond. In the crystal structure, a number of bound water molecules filling the tubes can be seen (Fig. 7c). Such tubes can be found in all the representatives of the TGF-β superfamily with known spatial structure. The tubes are extended along the contact between the $\alpha3$ helix and the β-sheets.

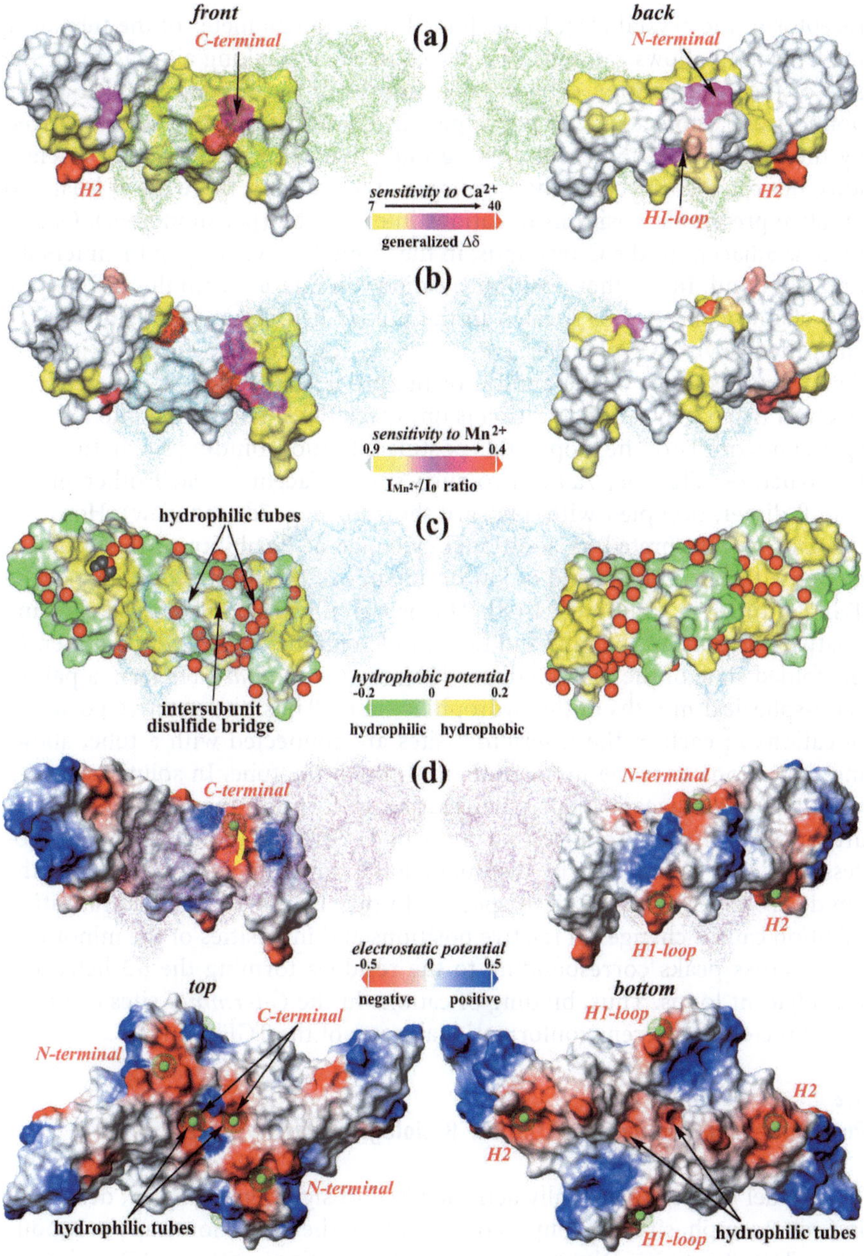

According to our prior data obtained for TGF-β3 [14] (Fig. 6c,d), the intersubunit interface and the $\alpha 3$ helix in particular are flexible and therefore the tubes are not always fully formed. The unfolded state of this part of the TGF-β3 dimer was observed for its complex with two ectodomains of the type II receptor in the crystal [16]. In the folded state, the diameter of the tube near the C-termini allows cations to be placed inside it, making allowance for the size of the coordination sphere. Polar groups of the protein, as well as the bound water molecules can participate in coordinating the cations. Judged by the locations of the carboxylate groups that favor coordinating calcium ions, but can also successfully coordinate other cations, there are at least two putative preferable positions of cations in each tube (per monomer). One of them is adjacent to the C-terminus, in the immediate vicinity of the intersubunit S–S bond. In another position, a cation can interact with the carboxylate side chain of the Asp 55 located in the pre-α3 helix loop and with the side chain hydroxyl oxygen of Ser 112 (Fig. 6e). According to our experimental data, the sensitivity of the amides of the pre-α3 helix loop residues, inclusively of Asp 55, to cation addition is unexpectedly small, taking into account spatial proximity of the loop to this postulated cation-binding site in the crystal structure (Figs. 6e, 7a,b). Two such sites adjacent to each other in the TGF-β dimer, occupied with ions, are shown in Fig. 7d (*top* view). However, this can be interpreted as additional evidence of flexibility of this loop together with the α3 helix and of low occupancy of the fully folded state of the TGF-β3 dimer in solution [14, 68]. This is well illustrated by Fig. 6d showing relative positions of the ions and the flexible regions of the TGF-β3 dimer. In the folded state of the TGF-β3 dimer [51], the *C-terminal* sites form a pair of semi-spherical mouths of the hydrophilic tubes. Thus, the putative positions of cations in each of the *C-terminal* sites are connected with a tube, allowing the cations to travel between them through the tube. In solution, the α3 helix and the adjacent loops participating in forming the hydrophilic tube are involved in slow (millisecond to second) conformational exchange process (Fig. 6c), as revealed by disproportional doubling and/or tailing of the amide cross-peaks in the NMR spectra. In the ^{1}H-^{15}N HSQC spectra, cation addition causes changes of relative positions and intensities of the minor and major cross-peaks corresponding to the residues forming the α3 helix and the adjacent loops. Thus, binding of cations in the *C-terminal* sites can alter occupancies of different conformational states of the TGF-β dimer.

3.2
Possibilities for Modulation of the TGF-β Biological Activity by Cations

The model of the dynamically activated TGF-β signaling complex, described in Sect. 2, emphasizes the physiological role of the found ion traps in modulating TGF-β activity. Direct interaction of the cations with the TGF-β ligand in each of the identified binding sites has a potential for differently affecting the biological activity of the protein. The *H2* binding sites located on the α2 helix surfaces in close proximity of both *wrist* and *knuckle/fingertips* epitopes for TβRI and TβRII, respectively, might play a role in modulation of the interaction of both epitopes with the receptors (Fig. 6d). The *N-terminal* and

H1-loop cation-binding sites can stabilize the flexible [14] α1 helix designated as the "*thumb*" epitope, which is spatially close to the *wrist* epitope (Fig. 6d). This segment is found only among the TGF-β isoforms and the activins but not in other members of the TGF-β superfamily and may also participate in the recognition of type I receptor or other TGF-β activity regulation proteins, such as betaglycan.

Other, *C-terminal* binding sites are remote from the receptor-binding epitopes, however, their location in the mobile dimer interface implies interesting possibilities for regulation of TGF-β properties and thus modulation of its functioning. The *C-terminal* cation-binding cavities are in close proximity of the intermonomer S–S bond ("pivotal point" of the dimer), and the highly flexible pre-α3 helix loops [14] of the *wrist* epitopes participate in their formation (Fig. 6d). Particularly, in each monomer the side chain of Asp 55 directed towards the dimer interface (according to the spatial structures of TGF-β isoforms in the stretched conformation, PDB codes 1tgj, 2tgi, 1klc) is capable of holding cations in the cavities. The Asp 55 was shown to be a "hot spot" residue, critical for activity of TGF-β isoforms [53]. An analogous pre-α3 helix loop in the BMP-2 was shown to be critical for binding to the type I receptor [54]. Thus, insertion of cations into cavities inside the dimer interface can directly affect the dimer flexibility and its conformations. For example, stabilization of the α3 helix and its pre-helix loop upon cation binding can explain the stabilizing effect of Ca^{2+} for TGF-β3 pharmaceutical compositions [43], since this may prevent deamidation of Asn 42 and Asn 103 adjacent to the α3 helix in the dimer interface and is possibly crucial for chemical degradation of TGF-β3. As opposed to the other sites, cation binding in the intersubunit interface can effectively modify global conformational properties of the dimer and hence control activation of the TGF-β signaling complex. A unique role of the cations in the *C-terminal* pockets in the light of the dynamical model of TGF-β signaling complex activation can be to guide the allowable conformational rearrangements during signaling complex formation, and at the final stage, when the TRβI epitope is assembled, to stabilize the fully extended conformation of the TGF-β ligand, in which the cavities for the ions are fully folded.

3.3 The Role of Ion Specificity and Ion Interplay in the Low-Affinity (and Low-Selectivity) Ion Traps in TGF-β Function

One characteristic common to all organisms is the dynamic ability to coordinate constantly one's activities with environmental changes [69]. In many cases, variations of a single parameter of a biological system cause complex responses in organisms. Any complex non-linear signaling system can be described in terms of a network with multiple nodes and junctions, where the incoming signals are converted, processed and then routed to multiple outputs. In order to understand the response of living organisms to the environmental changes, both detailed insights into the function and mechanism of each component of the signaling system and generic foresight of the overall network architecture are required.

As was pointed out above, cation binding to the TGF-β ligand can affect both local and global conformational preferences of TGF-β. Since all the cation-binding sites were shown to comparably interact with different ions, the regulation of signaling process by the environment composition can be complicated by the interplay between biologically significant ions. Physiological manifestations of the competitive interplay between Na^+, K^+, Ca^{2+}, Mg^{2+} and some other ions have been reported [23, 44, 70, 71]. Specifically to TGF-β, correlation between the Na^+/K^+ ratio, concentration of TGF-β2 and risk of malignant tumor development was also observed [72]. Moreover, metal debris from implants made of a much wider spectrum of metals, including those normally present in the organism and otherwise, has been shown to differentially affect osteoblast proliferation, differentiation and apoptosis, inclusively through changes in the levels of TGF-β1, tumor necrosis factor α, and a variety of other signaling proteins [73]. Such effects can be partly explained by active and passive transmembrane transport and interaction of the ions with ion-specific receptors, present in the cellular membranes. However, there is an integrated and closely interrelated system of feedbacks at every level of interaction of all the participants of the cell-signaling process, e.g. the changes in the levels of active TGF-β result in delayed modulation of TGF-β synthesis [2, 9]. Therefore, the observed direct interaction of TGF-β ligands with Na^+, Ca^{2+}, and Mn^{2+} ions suggests that chelation of the cations within the low-affinity ion traps can represent one of the regulation levels of TGF-β signaling.

As opposed to high-affinity, high-selectivity sites, such as EF-hand Ca^{2+}-binding proteins, low-affinity (and low-selectivity) ion traps allow combinatorial interaction with a certain spectrum of ions. Competition for the ion traps between different kinds of cations, including such biologically important ones as Ca^{2+} and Na^+, is another potential tier of the feedbacks. In particular, the competition of the Ca^{2+} and Na^+ ions, normally present in the extracellular environment in relatively high concentrations in different tissues of the organism (e.g. in vascular, renal, cardiac, and bone tissues), for the TGF-β ion traps could play a role in fine-tuning of the ligand behavior. Besides, such an important factor as pH of the environment can change cation affinities of the ion traps and cation interplay in them. This can be especially important for the processes of bone remodelling, which involves low-pH digestion of the bone by osteoclasts at the ruffled border [38, 39].

Thermodynamically, the role of cation binding can be described through redistribution of free energies of different conformations of the ligand, changing the kinetics and/or the spectrum of the allowable phase trajectories of ligand–receptor complex formation. The biological activity of various TGF-β representatives having special internal dynamics and structural properties would be regulated by different ions to different extents. The electrostatic contribution of calcium, as a divalent ion, into the free energy of the signaling complex formation is higher than that of sodium, given that their radii are very close. Besides that, calcium ions are known to have unique flexibility in the coordination numbers, ranging from 6 to 9 [22]. Therefore, calcium in the identified low-selectivity ion traps (especially in the *C-terminal* one) with comparable binding constants for calcium and sodium would have larger and potentially more diverse effects on the ligand behavior during receptor recognition and binding. Thus, local changes in calcium concentrations would affect structure and dynamical properties of TGF-β ligands to a greater extent, having larger consequences for physiological response to TGF-β signaling in organisms.

3.4 Regulation of TGF-β Activity by a Low-Affinity Sodium and Calcium Sensing Mechanism and Its Possible Physiological Consequences

Taking into account sodium's and calcium's unique role in cell-signaling networks, their role in modulating TGF-β activity deserves separate discussion. Mutations of TGF-β ligands, unlike mutations of the receptors and of the participants of the intracellular signaling cascade, have been shown to cause hypertension and osteoporosis [4, 6]—the diseases, directly associated with sodium and calcium metabolism. TGF-β ligands were also shown to play an important role in the development of the diseases [4, 6, 7, 74, 75]. Taking into account that hypertension may be associated with increased sodium concentration in blood [7, 9], it is tempting to assume that some of the mutations can affect modulation of the TGF-β activity by sodium concentration through sodium interaction with the low-affinity ion traps. Salt-sensitive hypertension causes further damage of other organs, primarily kidney, as well as vascular remodelling, and TGF-β is directly involved in the feedback system controlling development of the damage [9].

Extracellular Ca^{2+} concentrations are known to affect the balance between proliferation and differentiation in a number of cell types, including, e.g., breast epithelial cells, fibroblasts, osteoblasts and osteoclasts [23, 76]. Several mechanisms have been proposed for the effects of external Ca^{2+} concentrations upon cell-signaling systems. There is accumulating evidence suggesting a role for extracellular calcium in controlling cellular response to TGF-β [31–34]. However, there is no complete understanding yet of how calcium contributes to TGF-β signaling or what the mechanisms of calcium modulation are. For example, the calcium receptors involved in some of these mechanisms are activated by several-fold lower Ca^{2+} (and other divalent ions) concentrations compared to the concentrations modulating the functions of osteoclasts [23] and therefore these receptors cannot sense changes of Ca^{2+} concentration in the range of tens of millimoles. The existence of local extracellular calcium-sensing mechanisms acting independently of calcium receptors has been postulated by several authors on the basis of circumstantial evidence [23, 28]. This suggests that there are various different mechanisms involved in regulation of TGF-β activity by extracellular Ca^{2+}. Modulation of TGF-β activity through Ca^{2+} ion binding can represent an important regulation level of TGF-β signaling. This is consistent with the synergism between Ca^{2+} and TGF-β1, observed in [77, 78], where their co-administration produced greater than additive biological response on osteoblasts and osteolitic metastatic breast cancer cells. The need for existence of low-affinity sensing and regulation mechanisms is most obvious in the case of bone remodelling by osteoclasts. High concentration of extracellular Ca^{2+} is one of the main factors regulating bone remodelling, by means of a multi-organ cross-signaling cascade [38]. During the bone resorption by osteoclasts, Ca^{2+} and activated TGF-β are released simultaneously from the bone matrix. The release of the activated TGF-β can stimulate osteoblast proliferation and bone restoration [39, 75, 76], and chelation of the Ca^{2+} ions within the identified ion traps could modulate the conformational behavior of the TGF-β dimer and thus its biological activity, explaining the effect of certain mutations of TGF-β ligands for development of osteoporosis and other disturbances of bone metabolism.

4 Concluding Remarks

The last decade was characterized by discovery of many crucial events in TGF-β signaling at the molecular and cellular levels. The role of the TGF-β pathway in cellular homeostasis and control of cell fate has become increasingly clear. Sufficient biochemical and structural information about the TGF-β ligands and receptors was accumulated to propose a detailed dynamical model of active signaling complex formation. This is supplemented by identification of various signaling modes involving an extensive molecular toolkit, as well as their interaction and regulation. Expansion of the knowledge of the diverse and often seemingly contradictory TGF-β functions emphasizes the complexity of interactions and feedbacks between various participants of signal transduction, depending on the local environment, both extra- and intracellular. Identification of several cation-binding sites, or ion traps, some of which are located in the sensitive spots on the surface of TGF-β ligands, suggests another level of fine regulation of TGF-β activity by environmental parameters and of interdependence with other signaling systems. Interaction of physiologically relevant cations with these sites can affect the behavior of TGF-β during signal transduction in a manner, consistent with the available information on interrelation between the effects of TGF-β and ion concentrations in organisms. The sites have low affinity and low selectivity, allowing physiological concentrations of different cations to interdependently modulate TGF-β behavior. At the molecular level, the role of the ion traps in the TGF-β function is easily explainable by a model of the dynamically activated TGF-β signaling complex, assuming cooperative interaction of the TGF-β ligand with its receptors. The cation-binding sites located within the mobile dimer interface are capable of playing a unique role, controlling and fine-tuning conformational flexibility of the TGF-β dimer.

The progress in understanding TGF-β signaling roles and mechanisms resulted in applications in medicine and biotechnology, including artificial intervention into stem cell growth and differentiation, or development of new generations of pharmaceutical compositions for tissue restoration and treatment of such diseases as cancer, cardiovascular or renal dysfunctions, and immune disorders.

Acknowledgements This work was supported by grants from the Russian Foundation of Basic Research and the Ministry of Science and Education of the Russian Federation. E.V.B. thanks personally K.A. Beirit for financial support.

References

1. Massagué J (1998) Annu Rev Biochem 67:753
2. Massagué J, Chen Y-G (2000) Genes Dev 14:627
3. Massagué J, Blain SW, Roger SL (2000) Cell 103:295
4. Blobe GC, Schiemann WP, Lodish HF (2000) N Engl J Med 342:1350
5. Letterio JJ, Roberts AB (1998) Annu Rev Immunol 16:137
6. Miyazono K, Kusanagi K, Inoue H (2001) J Cell Physiol 103:295
7. Lijnen PJ, Petrov VV, Fagard RH (2003) Am J Hypertens 16:605
8. Reeves WB, Andreoli TE (2000) Proc Natl Acad Sci USA 97:7667
9. Sanders PW (2004) Hypertension 43:142
10. Sharma K, McGowan TA (2000) Cytokine Growth Factor Rev 11:115
11. Katagiri T, Takahashi N (2002) Oral Dis 8:147
12. Shi Y, Massagué J (2003) Cell 113:685
13. Piek E, Heldin C-H, ten Dijke P (1999) FASEB J 13:2105
14. Bocharov EV, Korzhnev DM, Blommers MJJ, Arvinte T, Orekhov VY, Billeter M, Arseniev AS (2002) J Biol Chem 277:46273
15. Greenwald J, Vega ME, Allendorph GP, Fischer WH, Vale W, Choe S (2004) Mol Cell 15:485
16. Hart PJ, Deep S, Taylor AB, Shu Z, Hinck CS, Hinck AP (2002) Nat Struct Biol 9:203
17. Thompson TB, Woodruff TK, Jardetzky TS (2003) EMBO J 22:1555
18. Souchelnytskyi S, Moustakas A, Heldin C-H (2002) Trends Cell Biol 12:304
19. Derynck R, Feng X-H (1997) Biochim Biophys Acta 1333:105
20. Blobe GC, Schiemann WP, Pepin M-C, Beauchemin M, Moustakas A, Lodish HF, O'Connor-McCourt MD (2001) J Biol Chem 276:24627
21. Qin BY, Lam SS, Correia JJ, Lin K (2002) Genes Dev 16:1950
22. Campbell AK (1983) Intracellular calcium: Its universal role as regulator. John Wiley & Sons Ltd., Norwich
23. Brown EM, MacLeod RJ (2001) Physiol Rev 81:239

24. Raya Á, Kawakaml Y, Rodriguez-Esteban C, Ibanes M, Rasskin-Gutman D, Rodriguez-León J, Büscher D, Feijó JA, Belmonte JCI (2004) Nature 427:121
25. Berridge MJ, Bootman MD, Roderick HL (2003) Nat Rev Mol Cell Biol 4:517
26. Webb SH, Moreau M, Leclerc C, Miller AL (2005) Cell Calcium 37:375
27. Ishiyama N, Shibata H, Kanzaki M, Shiozaki S, Miyazaki J-I, Kobayashi I, Kojima I (1996) Mol Cell Endocrinol 117:1
28. McGowan TA, Madesh M, Zhu Y, Wang L, Russo M, Deelman L, Henning R, Joseph S, Hajnóczky G, Sharma K (2002) Am J Physiol 282:F910
29. Sharma K, McGowan TA, Wang L, Madesh M, Kaspar V, Szalai G, Thomas AP, Hajnóczky G (2000) Am J Physiol 278:F1022
30. Chen C-H, Seguin-Devaux C, Burke NA, Oriss TB, Watkins SC, Clipstone N, Ray A (2003) J Exp Med 197:1689
31. Martineau I, Lacoste E, Gagnon G (2004) Biomaterials 25:4489
32. Nakade O, Takahashi K, Takuma T, Aoki T, Kaku T (2001) J Bone Miner Metab 19:13
33. Wang D, Canaff L, Davidson D, Carluka A, Liu H, Hendy GN, Henderson JE (2001) J Biol Chem 276:33995
34. Sakaguchi M, Sonegawa H, Nukui T, Sakaguchi Y, Miyazaki M, Namba M, Huh N (2005) Proc Natl Acad Sci USA 102:13921
35. Muldoon LL, Rodland KD, Magun BE (1988) J Biol Chem 263:18834
36. Gooch JL, Gorin Y, Zhang B-X, Abboud HE (2004) J Biol Chem 279:15561
37. Chuang C-C, Tan S-K, Tai L-K, Hsin J-P, Wang F-F (1998) Mol Pharmacol 53:1054
38. Purroy J, Spurr NK (2002) Hum Mol Genet 11:2377
39. Mostov K, Werb Z (1997) Science 276:219
40. Gombotz WR, Pankey SC, Bouchard LS, Phan DH, Puolakkainen PA (1994) J Appl Biomater 5:141
41. Blom EJ, Klein-Nulend J, Kurashina K, van Waas MA, Burger EH (2000) J Biomed Mater Res 50:67
42. Kamakura S, Sasano Y, Nakajo S, Shimizo T, Suzuki O, Katou F, Kagayama M, Motegi K (2001) J Biomed Mater Res 57:175
43. Arvinte T, Schote UT, Sigg J (2000) Patent WO/2000/054797
44. Quinn SJ, Kifor O, Trivedi S, Diaz R, Vassilev P, Brown E (1998) J Biol Chem 273:19579
45. Ying W-Z, Sanders PW (2002) Hypertension 39:239
46. Murphy-Ullrich JE, Poczatek M (2000) Cytokine Growth Factor Rev 11:59
47. Mangasser-Stephan K, Gressner AM (1999) Cell Tissue Res 297:363
48. Ten Dijke P, Hill CS (2004) Trends Biochem Sci 29:265
49. Sebald W, Mueller TD (2003) Trends Biochem Sci 28:518
50. Derynck R, Zhang YE (2003) Nature 425:577
51. Mittl PRE, Priestle JP, Cox DA, McMaster G, Cerletti N, Grütter MG (1996) Protein Sci 5:1261
52. Kirsch T, Sebald W, Dreyer MK (2000) Nat Struct Biol 7:492
53. Huang SS, Zhou M, Johnson FE, Shieh H-S, Huang JS (1999) J Biol Chem 274:27754
54. Keller S, Nickel J, Zhang J-L, Sebald W, Mueller TD (2004) Nat Struct Mol Biol 11:481
55. Greenwald J, Groppe J, Gray P, Wiater E, Kwiatkowski W, Vale W, Choe S (2003) Mol Cell 11:605
56. Hinck AP, Archer SJ, Qian SW, Roberts AB, Sporn MB, Weatherbee JA, Tsang ML-S, Lucas R, Zhang B-L, Wenker J, Torchia DA (1996) Biochemistry 35:8517
57. Pellaud J, Schote U, Arvinte T, Seelig J (1999) J Biol Chem 274:7699
58. Keah HH, Hearn TW (2005) J Mol Recognit 18:385
59. De Creszenso G, Grothe S, Zwaagstra J, Tsang M, O'Connor-McCourt MD (2001) J Biol Chem 276:29632

60. De Creszenso G, Pham LP, Durocher Y, O'Connor-McCourt MD (2003) J Mol Biol 328:1173
61. De Creszenso G, Pham LP, Durocher Y, Chao H, O'Connor-McCourt MD (2004) J Biol Chem 279:26013
62. Moriki T, Maruyama H, Maruyama IN (2001) J Mol Biol 311:1011
63. Chi J-I, Yokota H, Kim S-H (1997) FEBS Lett 414:327
64. Bell CA, Tynan JA, Kristen CH, Meyer AN, Robertson SC, Donoghue DJ (2000) Mol Biol Cell 11:3589
65. Fleishman SJ, Schlessinger J, Ben-Tal N (2002) Proc Natl Acad Sci USA 99:15937
66. Zhu H-J, Sizeland AM (1999) J Biol Chem 274:11773
67. Zhu H-J, Sizeland AM (1999) J Biol Chem 274:29220
68. Bocharov EV, Blommers MJJ, Kuhla J, Arvinte T, Bürgi R, Arseniev AS (2000) J Biomol NMR 16:179
69. Jordan JD, Landau EM, Iyengar R (2000) Cell 103:193
70. Korolev N, Lyubartsev AP, Rupprecht A, Nordenskiöld L (1999) Biophys J 77:2736
71. Di Paolo ML, Scarpa M, Corazza A, Stevanato R, Rigo A (2002) Biophys J 83:2231
72. Erbas H, Lennard TW, Lai LC (1999) Anticancer Res 19:3077
73. Hallab NJ, Vermes C, Messina C, Roebuck KA, Glant TT, Jacobs JJ (2002) J Biomed Mater Res 60:420
74. Eisman JA (1999) Endocrine Rev 20:788
75. Janssens K, ten Dijke P, Janssens S, Hul WV (2005) Endocrine Rev 26:743
76. Clines GA, Guise TA (2005) Endocrine-Related Cancer 12:549
77. Jaunberzins A, Gutmann JL, Witherspoon DE, Harper RP (2000) J Endodont 26:494
78. Sanders JL, Chattopadhyay N, Kifor O, Yamaguchi T, Butters RR, Brown EM (2000) Endocrinology 141:4357
79. Efremov RG, Vergoten G (1995) Phys Chem 99:10658
80. Koradi R, Billeter M, Wüthrich K (1996) J Mol Graphics 14:51

Top Curr Chem (2008) 273: 183–202
DOI 10.1007/128_2007_19

Published online: 29 February 2008

Molecular Recognition of Ligands by Native Viruses and Virus-Like Particles as Studied by NMR Experiments

Christoph Rademacher · Thomas Peters (✉)

Institute of Chemistry, University of Luebeck, Ratzeburger Allee 160, 23538 Luebeck, Germany
thomas.peters@chemie.uni-luebeck.de

Abstract Viral entry into host cells is a process that in the majority of cases is not understood in its molecular details. The first step of viral entry is the recognition of cellular receptors on host cells by viruses, and although X-ray crystallography had yielded some spectacular results in individual cases, in general there is little data available to unravel the principles of virus–ligand recognition at atomic resolution. Therefore, new techniques that uncover the molecular details of these recognition processes are needed. The investigation of virus–ligand interactions using ligand-based NMR techniques is an emerging field with the potential to substantially contribute to a deeper understanding of the molecular aspects of viral entry into host cells. Here, we give an overview that covers some of the systems studied so far. This comprises native viruses as well as virus-like particles (VLPs). We will not address studies that have been performed with individual proteins that are not in a native environment. It turns out that STD NMR in particular has a great potential to shine light on the viral entry process as this technique requires only very moderate amounts of viruses or VLPs and corresponding ligands. As a further advantage, this approach is also applicable to ligands that bind to viruses with medium to low affinity. Therefore, STD NMR is extremely well suited for development of antiviral entry inhibitors utilizing fragment-based approaches with low molecular weight compounds.

Keywords Entry inhibitors · Ligand-based NMR · Noroviruses · Rabbit hemorrhagic disease virus · Rhinoviruses · STD NMR · Virus–ligand interaction

1 Introduction

The first step during a viral infection is the entry of a virus into the host cell. Viral entry is triggered by specific interactions between viral coat or envelope proteins and one or more cell surface receptors. In the field of virology, much effort has been devoted to the identification of receptors and attachment factors [1–5]. Yet, only limited knowledge is available about the molecular recognition mechanisms during attachment at atomic resolution [1]. The cell tropism critically depends on the fine details of this attachment process. Any attempt to prevent viral entry requires precise knowledge about the specific interactions between viral surface proteins and cellular receptors [6]. It is observed that small changes in amino acid composition of the viral surface proteins involved in the recognition process often lead to significant changes in binding specificities, and therefore facilitate the crossing of species barriers [2, 4, 7, 8].

The preference for a certain receptor is diverse even among members of the same virus family. Additionally, host cell tropism fluctuates between members of the same genus and even in isolates of one virus [1, 4, 8]. Knowledge of the molecular recognition of receptors by viruses at atomic resolution is limited and relies mostly on demanding crystallographic studies. Almost no information about dynamic contributions to the recognition process is available.

Viral entry into the host cell can be divided into three steps: receptor binding, internalization, and post-internalization trafficking [2, 9]. Insight into the molecular details of the recognition process will facilitate the understanding of the systems biology of virus entry [6]. Two major routes are taken by viruses to enter host cells, the endocytotic (e.g. HRV) and the non-endocytotic pathway (e.g. HIV). In either case a significant conformational change in the virus attachment factor, the host cell receptor, or both is required if a high binding affinity is encountered. Low affinity binding of low molecular weight ligands, (e.g. sialic acid residues) is associated with less conformational changes [1].

Although non-enveloped viruses have a relatively simple structure compared to enveloped viruses, only few examples of structural rearrangements upon receptor binding have been investigated in detail [10–12]. A metastable state is of significant importance during viral entry and has to fulfill two needs. First, the membrane of the host cell has to be destabilized; second, the genome has to be transferred into the cytoplasm. For this latter step clearly major conformational changes have to be induced [10]. Interestingly, these alterations in the viral shell can be induced by soluble receptors or receptor fragments. This may offer an opportunity to destabilize the viruses prior to host cell entry as the metastable state is not reversible [13]. Other antiviral strategies try to prevent the viral capsid from entering the metastable state. The "breathing" of the virus is impeded and thereby the infectious genome is not released into the cell. This strategy has been successfully applied to human rhinoviruses (HRV) [14–16]. The effect of the highly flexible ligands on prevention of the transition is proposed to be entropic in nature [17, 18]. Therefore, knowledge of the bioactive conformation is of significant importance.

An attractive option for the development of antiviral compounds is to prevent viral entry. These entry inhibitors interfere at an early stage of infection, which has several advantages. The number of infectious particles is rather small, and therefore entry inhibitors should be very effective if applied at the right time, even at lower doses [2]. In addition, many of the problems associated with delivering a drug into cells are avoided as the interference takes place outside the cell.

So far, only few entry inhibitors are used in antiviral therapies [2]. Synagis and T-20 are the only entry inhibitors on the market. Synagis (palivizumab) is a monoclonal antibody that blocks the attachment of the respiratory syncytial virus. The fusion of HIV with its host cell is inhibited by T-20 (enfuvirtide). T-20 and the recently described VIRIP are larger peptides and excellent examples, where knowledge of the bioactive conformation plays a key role in drug development. VIRIP is a natural peptide that blocks HIV entry in a similar fashion to enfuvirtide, with the important difference that this new peptide is significantly smaller and interacts with a different region of gp41 [19]. A number of attempts in anti-HIV drug development are focusing on the search for entry inhibitors [19–21]. But many obstacles still have to be resolved during development of entry inhibitors. Escape mutants and viruses that avoid transport through the extracellular milieu via transcytosis [22] are examples of hurdles that have to be overcome.

The investigation of viral entry at atomic resolution is mainly based on data from X-ray crystallography or from cryo-electron microscopy. A comprehensive database covering the structures of viruses is available on the web (ViperDB) [23, 24]. With increasing size of the virus it becomes more difficult to yield high resolution crystallographic structures [25]. The combination of cryo-electron microscopy and X-ray crystallography has led to recent progress in studying large biological assemblies [26]. Yet the crystallization of viruses or virus-like particles in complex with their receptor, or fragments thereof, is far from being trivial [25, 27]. Therefore, structural data on receptor/virus complexes are sparse [28–30].

Albeit a high specificity of receptor recognition, affinities of individual receptor fragments are often rather low [1, 5]. However, as receptors form clusters on host cell surfaces, the avidity increases. The low affinity of individual ligands lends itself well to the application of ligand-based NMR techniques.

We will summarize studies that have been performed to characterize the interaction of small molecule ligands with viruses and virus-like particles (VLP) by NMR spectroscopy. It has to be noted that, strictly speaking in terms of viral attachment, the virus represents the ligand binding to a receptor or receptor fragments. Here, the term "ligand" will be used more intuitively for a small molecule binding to a virus, whereas the virus or VLP will be called "receptor" throughout this communication.

For the analysis of viruses recognizing low molecular weight compounds, one relies on the application of ligand-based techniques, whereas receptor-based experiments are hampered by the size of the viral capsid. There have been recent attempts to characterize viral assembly with protein-based NMR techniques [31], but they are not suitable to monitor the interaction between an intact capsid and a small molecule. To our knowledge there is no example in the literature where single structural proteins of a virus are sufficient for internalization. Only whole virus particles can enter the host cell. Therefore, it is of considerable interest to map the binding of ligands to viruses utilizing ligand-based NMR techniques. It will be shown below that the large size of the receptor (virus) has important consequences for the experimental design.

2 NMR Studies of Virus–Ligand Interactions

2.1 NMR Studies with Influenza Viruses

National response plans and stockpiling of antiviral drugs reflect the urgent need for the development of flexible, innovative treatments against Influenza virus infections. Strategies would be desirable that allow for a fast screening of virus–receptor interactions in order to facilitate access to tailored entry-inhibitors targeting newly surfacing viral mutants.

These demands arise because the Influenza virus genome underlies considerable antigenic variation. An animal reservoir combined with a high mutation rate is an excellent prerequisite for its pandemic occurrence [32, 33]. Only a few single point mutation sites in the hemagglutinin are responsible for a switch between the preference for α-(2,6)- over α-(2,3)-linked sialic acids as a receptor determinant. It is believed that this change in specificity enables the virus to infect humans [34, 35], and therefore represents a key step in crossing the species barrier. Additionally, more than 40 viruses are known that bind to sialic acids [36], which renders the sialic acids as an attractive target for drug design in more general terms.

In 1992, Wiley and coworkers were the first to investigate ligand binding to native viruses by NMR [37]. Hemagglutinin (HA) and neuraminidase are the most prominent glycoproteins present on the viral surface and have been the target of various antiviral strategies [38, 39]. Natural and synthetic ligands were employed to monitor the binding to HA of H3N2 Influenza virus (Scheme 1). Binding constants were derived from the observed line broadening of ligand resonances.

Two strategies are followed in this study. In the absence of a compound with a known dissociation constant, an analysis of the line broadening of the ligand upon titration into a highly concentrated virus solution is performed. The ligand is in chemical exchange with a large receptor (virus) that imprints its significantly different relaxation properties onto the ligand. Depending on the kinetics of the exchange reaction and the excess of ligand over receptor, the average relaxation times of the ligand protons are altered upon binding. The observable factor in this experiment is the line broadening of ligand resonances in a simple ^{1}H spectrum. With a fast exchanging ligand and a small fraction of bound ligand molecules compared to the total concentration of ligand, conditions are met that allow usage of Eq. 1 in order to derive dissociation constants K_D:

$$[L]_{tot} = \frac{[P]_{tot}}{\Delta\Delta\upsilon}\Delta\Delta\upsilon_{Bapp} - K_D \tag{1}$$

with $[L]_{tot}$ and $[P]_{tot}$ being the total concentrations of the ligand and the protein (virus) respectively, whereas $\Delta\Delta\upsilon$ and $\Delta\Delta\upsilon_{Bapp}$ being the observed line broadening and the apparent broadening in the fully bound state of the ligand [40] (Fig. 1).

The second strategy is employed if a ligand with a known dissociation constant is available. This ligand is then used as a so-called "reporter ligand". A competition NMR experiment is performed in which the concentration of a ligand with unknown affinity is varied while keeping the concentration of the "reporter ligand" constant. The decreasing line broadening of a "reporter ligand" NMR signal is monitored while the ligand of unknown affinity is titrated (Fig. 2). Alternatively, the concentration of the "reporter ligand" is changed during the titration, and the concentration of the ligand of unknown affinity is kept constant.

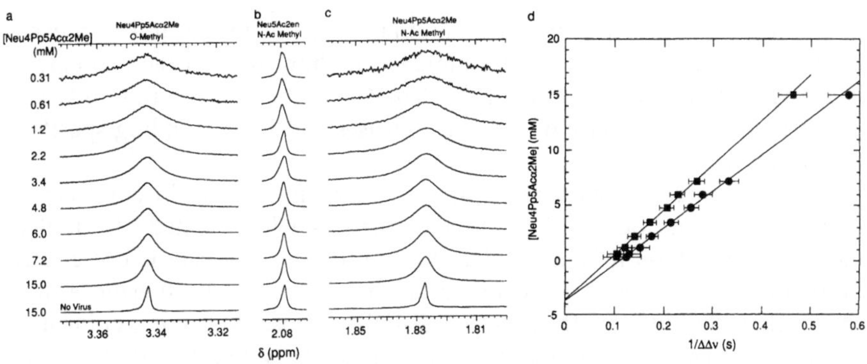

Fig. 1 ^{1}H NMR titration of Neu4PpAcα2Me in the presence of 3 mg mL^{-1} Influenza (H3N2) virus [37]. Two ligand resonances are observed: (*a*) *O*-methyl and (*b*) *N*-Ac methyl. The *N*-Ac methyl resonance of the neuraminidase inhibitor Neu5Ac2en (*c*) was used as a reference to calculate $\Delta\Delta\nu = (\Delta\nu_{Neu4PpAc\alpha2Me} - \Delta\nu_{Neu5Ac2en})_{virus} - (\Delta\nu_{Neu4PpAc\alpha2Me} - \Delta\nu_{Neu5Ac2en})_{no\ virus}$. *d* Linear regression analysis of titration of *N*-Ac methyl (•) and 2-*O*-methyl of Neu4PpAcα2Me (■), which yield – K_D (*y*-intercept). The error in broadening was estimated using the square root of the signal-to-noise ratio or the digital resolution, whichever was higher. The data were acquired at 500 MHz. (With permission of Virology.)

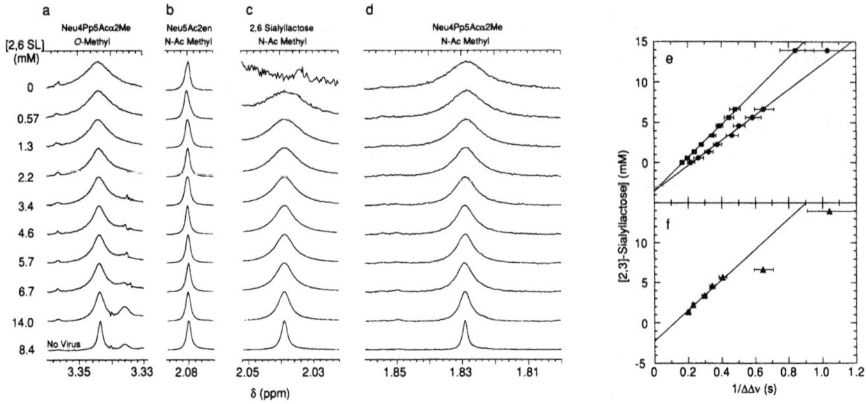

Fig. 2 ^{1}H NMR titration of α(2,6)sialyllactose using Neu4Pp5Acα2Me as a reporter ligand [37]. While Neu4Pp5Acα2Me was present at a constant concentration of 1 mM, α(2,6)sialyllactose was titrated into the solution in presence of the 2 mg mL^{-1} H3N2 Influenza virus. A decrease in line broadening of the *O*-methyl (*a*) and *N*-Ac methyl (*d*) resonances of Neu4Pp5Acα2Me can be observed when the concentration of α(2,6)sialyllactose is increased. To calculate $\Delta\Delta\nu$, the *N*-Ac methyl resonance of the Neu5Ac2en (*b*) is used. *c* *N*-Ac methyl resonances of α(2,6)sialyllactose. Linear regression analysis of *e* *N*-Ac methyl (•) and 2-*O*-methyl (■) of Neu4Pp5Acα2Me, and *f* 2-*O*-methyl (▲) of α(2,6)sialyllactose. The known $K_{D(\mathrm{Neu4Pp5Ac}\alpha\mathrm{2Me})} = 3.6$ mM of the reporter ligand was used to derive $K_{D(\alpha(2,6)\mathrm{sialyllactose})}$ by following the equation: *y*-intercept $= -K_{D(\alpha(2,6)\mathrm{sialyllactose})}(1 + [\mathrm{Neu4Pp5Ac}\alpha\mathrm{2Me}]/K_{D(\mathrm{Neu4Pp5Ac}\alpha\mathrm{2Me})})$. (With permission of Virology.)

In order to exclusively measure binding of ligands to HA and to prevent binding to neuraminidase, a neuraminidase inhibitor Neu5Ac2en (Scheme 1) is added at high concentrations (10–20 mM) in all cases. Neuraminidase is also present on the surface of Influenza viruses and shares binding specificities with HA.

Dissociation constants derived by these experiments are in good agreement with results obtained by similar analysis with isolated HA molecules [40]. Therefore, it is concluded that the HA binding site is not altered by removing the protein from its natural environment. Furthermore, it is shown that the binding affinity of α-2,6-sialyl lactose does not significantly differ from that of the simple monosaccharide derivative Neu4Pp5Acα2Me (Scheme 1). However, inversion of the anomeric center of the sialic acid residue leads to a total loss of binding affinity. For these studies, concentrations of 2–4 mg mL^{-1} of virus have been used.

It is clear that this approach does not deliver information about the binding process at atomic resolution. Rather, it focuses on the binding affinities of ligands to the virus. Careful data analysis allows discrimination of residues that actively participate in the binding from those that contribute less to the interaction. Ligands in the intermediate exchange regime on the chemical

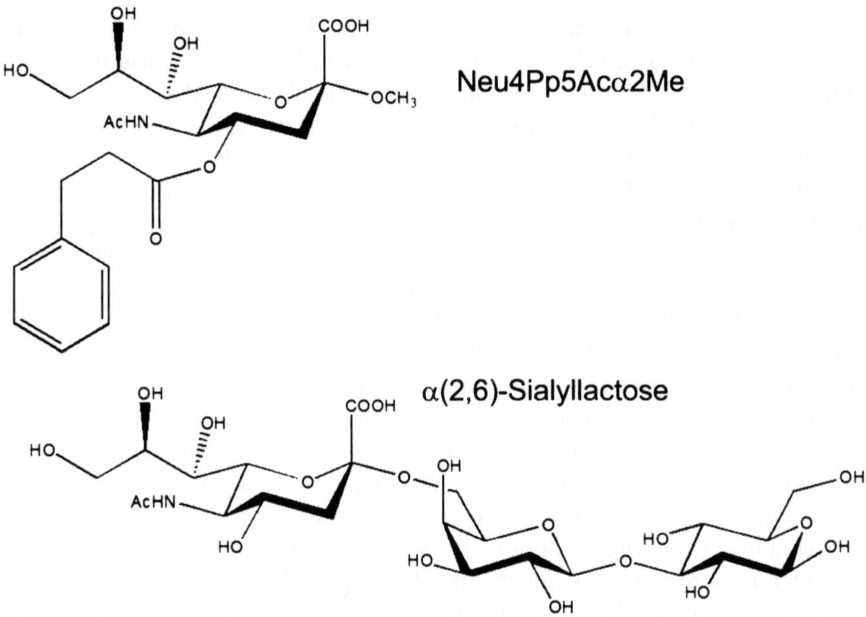

Scheme 1 Ligands used in titration studies of Wiley et al. [37]

shift timescale allow extending the analysis of data to extract potential multiple binding modes [41].

2.2 NMR Studies with Human Rhinoviruses

Human rhinoviruses (HRV) are members of the *Picornaviridae* family. The HRVs are classified according to their receptor specificity into members of the major and minor groups. The 87 members of the major-group viruses bind to the intracellular adhesion molecule receptor 1 (ICAM-1), whereas the 12 serotypes of the minor group bind to members of the low-density lipoprotein receptor family (LDLR) [42]. Rhinoviruses cause more than a billion cases of the common cold each year and are also associated with asthma exacerbations [43, 44]. Statistically, one encounters one to three infections per year on the average [45]. As a result, rhinoviral infections are responsible for 25 million days of missed work in the USA [46].

HRVs are non-enveloped viruses of icosahedral overall shape [44]. Located on the exterior of the viral capsid are three structural proteins (VP1, VP2 and VP3), each consisting of an eight-stranded antiparallel β-barrel. VP4 is found at the interface with the RNA inside the virus. A pocket factor is usually bound to a hydrophobic canyon binding site within the VP1 β-barrel. This lipid-like molecule is important for the stability of the capsid and has been the target of antiviral strategies [47]. Compounds have been designed and synthesized that specifically bind into the canyon binding site, leading to an inhibition of the "breathing" of the viral capsid. Binding of these so called WIN compounds blocks the release of RNA by locking VP4 inside the capsid and thus disabling transition to the metastable state [17, 48–52].

Utilizing STD NMR experiments we have resolved the binding epitope for one of the WIN compounds, REPLA 394, at atomic resolution [53]. Competitive STD NMR titrations then yielded the dissociation constant. When bound to native human rhinovirus serotype 2 (HRV2), saturation is predominantly transferred to the isoxazole and the benzoic acid residues, whereas the aliphatic linker receives less saturation (Fig. 3). Results reported before [16] are in accordance with the data reported in our study.

For our experiments, concentrations of 0.02 μM (85 μg) of HRV2 and not more than 0.45 mM (80 μg) of ligand were used in one NMR sample. Thus, the quantities of virus and ligand required for these kinds of experiments were significantly reduced when compared to the ^{1}H NMR experiments described above [37]. The amount of virus was two orders of magnitude lower, and ligand quantities were approximately 50 times less. In addition to binding affinities that are also available from simple ^{1}H NMR titrations, STD NMR also delivers binding epitopes at atomic resolution.

Furthermore, this study shows that the identification of binding ligands from compound mixtures is possible, also in the case of large receptor proteins such as a virus. Specifically, it was possible to unambiguously detect the binding of REPLA 394 to HRV2 even in the presence of a tenfold molar excess of methanol and *O*-methyl α-D-glucopyranoside [53]. To discriminate the binding of REPLA 394 from unspecific binding to hydrophobic pockets on the viral shell, a titration experiment was conducted in a similar fashion to

Fig. 3 Repla 394 binding to HRV2 as monitored by STD NMR. **a** Wire frame model of HRV2. The buried pocket factor is bound around the fivefold axis of the virus and is highlighted in *red* CPK representation. **b** STD NMR spectra of REPLA 394 binding to HRV2 (taken from [53]): reference spectrum of REPLA 394 are shown in presence (*i*) and in absence of HRV2 (*ii*). Asterisks (*) mark resonances of dimethylformamide resulting from the synthesis of the ligand. STD NMR spectra of HRV2 are shown in absence (*iii*) and presence (*iv*) of REPLA 394. Polyamines serve as counterions for the viral RNA and show STD effects as well as Repla 394. Dimethylformamide does not bind. **c** Scheme of the pocket factor bound to HRV2. This is based crystallographic data available for HRV2 (pdbcode: 1FPN) [93] and was produced using PoseViewWeb [94]. The hydrophobic cleft surrounding the aliphatic chain of the pocket factor is represented in *green*. Explicit interactions are found between the carboxylic group of the pocket factor and L1011 and Q1021. **d** Scheme of REPLA 394 bound to HRV2. This is a result of molecular docking using FlexX [95] in combination with the consensus scoring of Xscore [96]. The interaction of the carboxylic group with the canyon is conserved. A group epitope mapping according to Mayer and Meyer [97] is indicated by STD effects that are normalized to the highest STD observed for a ligand proton ▶

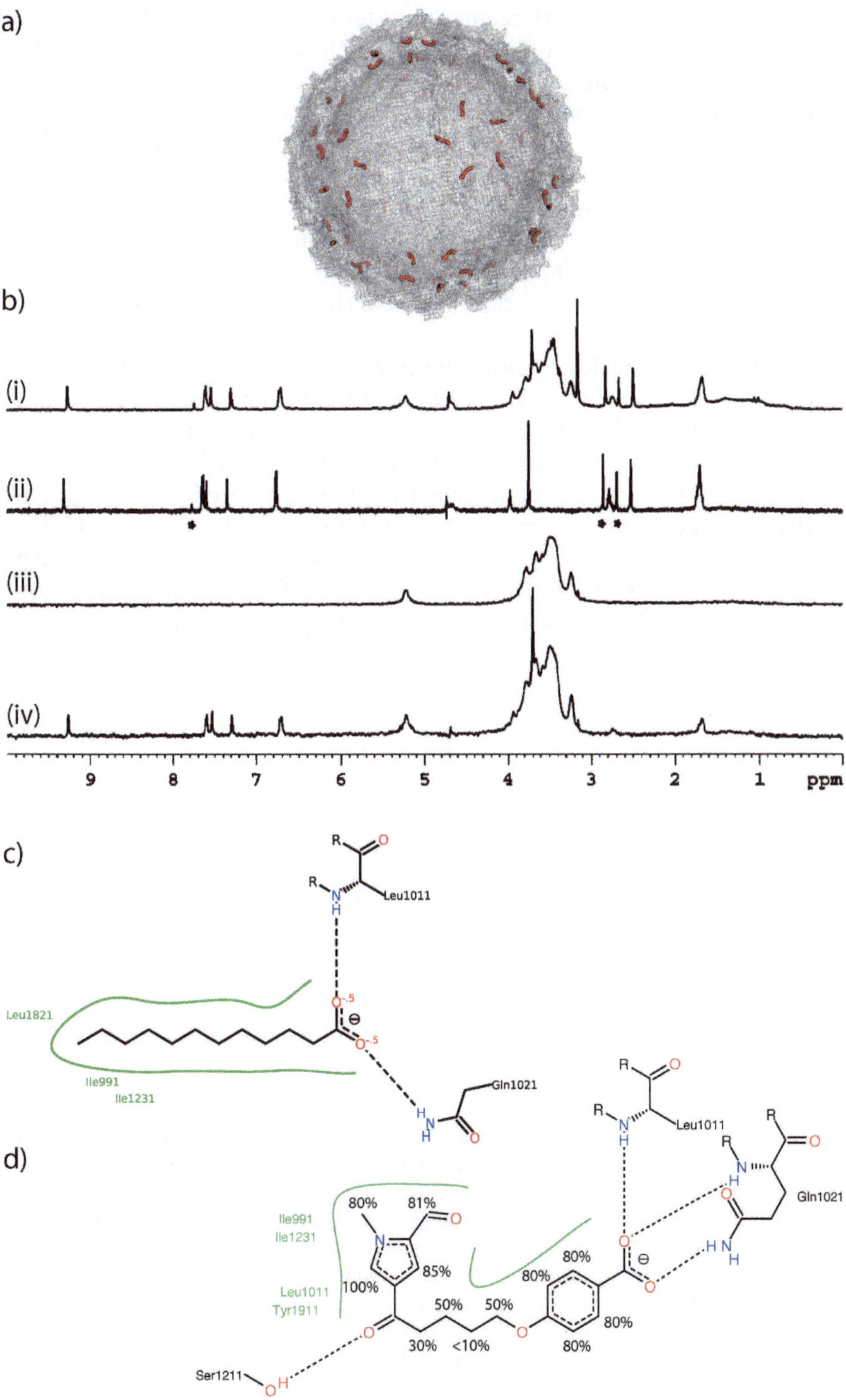
a)
b)
(i)
(ii)
(iii)
(iv)
9
8
7
6
5
4
3
2
1
ppm
c)
Leu1011
Leu1821
Ile991
Ile1231
Gln1021
d)
Leu1011
Gln1021
Ile991
Ile1231
Leu1011
Tyr1911
Ser1211
80%
81%
85%
100%
50%
50%
30%
<10%
80%
80%
80%
80%

that described above [37]. Here, the "reporter ligand" is WIN 52084, a ligand that is known to specifically bind to the hydrophobic canyon binding site [16]. Applying STD titration experiments, a dissociation constant was estimated. Finally, heat denaturation of the virus led to a loss of STD signals from REPLA 394, indicating that this compounds binds to the native virus but not to isolated capsid proteins. Therefore, this particular combination of STD NMR experiments clearly shows that binding of REPLA 394 is specific to the hydrophobic canyon that is present only in intact virus particles. This study also emphasizes the fact that in many cases it will be necessary to use intact viral particles instead of individual capsid proteins in order to study the binding of ligand molecules.

Besides inhibiting RNA release by making the viral capsid more rigid, blocking viral attachment to the host cell surface is an attractive target during the viral life cycle. The binding of members of the major group to host cells via ICAM-1 has been investigated in detail [54]. Soluble ICAM-1 receptor fragments were used to reduce infectivity [55]. In a recent crystallographic study, the structure of a member of the minor group rhinoviruses, HRV2, in complex with a receptor fragment of the very low-density lipoprotein receptor (VLDLR) was solved at 3.5 Å resolution [28]. This is in accordance with a previous study [56] that suggested that the receptor fragment V3 of VLDLR

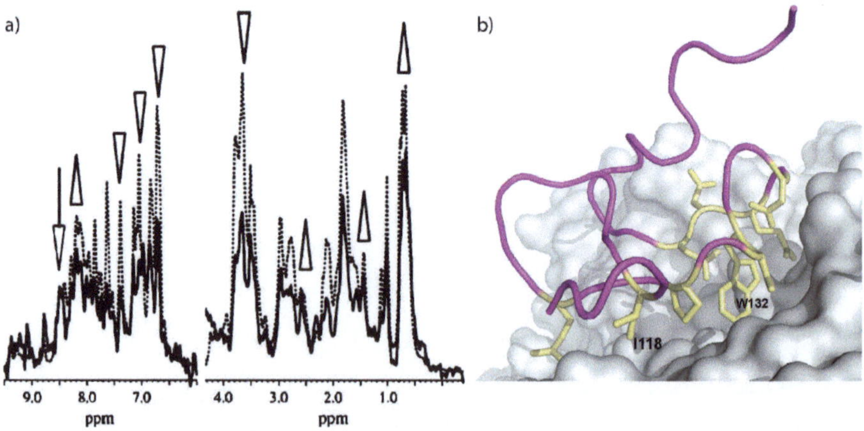

Fig. 4 Third receptor fragment V3 of VLDLR binding to HRV2 as monitored by STD NMR. **a** Dotted reference spectrum is overlaid with the corresponding STD NMR spectrum (taken from [57]). This data was recorded on a 500 MHz spectrometer with 500-fold excess of V3 over HRV2. *Up-arrows* highlight resonances with a large relative saturation transfer. *down-arrows* indicate resonances that received less relative saturation. **b** The complex as derived from a cocrystal structure (pdbcode: 1V9U) [28]. Amino acid side chains from V3 closer than 5.0 Å to the virus (*white*) are highlighted in *yellow*. The only aromatic amino acid close to the viral shell is W132 and the only methyl group in a 5.0 Å radius around the virus originates from I118. It is hypothesized that the observed saturation transfer is mainly due to these two amino acids

binds to HRV2 around its fivefold axis. From the crystallographic study, the protein–protein interaction surface between HRV2 and V3 of VLDLR is known [28]. This paves the way for ligand-based NMR techniques in order to map the binding of V3 to HRV2 at atomic resolution in solution. Moreover, it will be of interest to systematically monitor the interaction of other rhinoviruses with the receptor fragment, or with mutants thereof, using STD NMR experiments. First STD experiments with V3 and HRV2 showed promising results [57] (Fig. 4). Finally, it will interesting to compare the interaction of HRV2 with VLDLR to the binding of "natural ligands" (e.g., apoE) to VLDLR [58, 59].

2.3 Rabbit Hemorrhagic Disease Virus

Blood group determinants are carbohydrate structures and are involved in various cellular and immunological processes such as cancer, reproduction, and apoptosis [60–62]. Interestingly, histo-blood group antigens (HBGAs) are also recognized by a variety of pathogens. A high degree of genetic polymorphism in glycosyltransferase genes determines the exposition of HBGAs on human cells, and helps against invading pathogens [63]. This mechanism builds up a herd immunity in the population to protect individuals against infectious diseases [60, 64]. HBGAs cover a large surface of the body facing the environment (e.g., epithelial cells covering lung and intestine tissue) and hence involve intense contact with microorganisms.

One animal pathogen that binds to HBGAs is the rabbit hemorrhagic disease virus (RHDV) [60, 65]. For rabbits the polymorphism in HBGAs is not known, which amongst others is the reason for its very high lethality [66]. RHDV is a member of the Lagovirus genera of the *Caliciviridae* family. Its first occurrence was described in 1984, and studies suggest that its pathogenicity developed a few years before that [66, 67]. Thus, RHDV is an example of an emerging viral threat. Crossing the species barrier is a common feature for *Caliciviridae* [68, 69].

Members of the *Caliciviridae* family can hardly be examined in cell culture or animal models. Therefore, so-called virus-like particles (VLP) are employed in current experiments. These particles are expressed recombinantly in insect cells using a baculovirus system and do not carry infectious viral RNA [70–72]. It has been shown by single particle tracking studies that VLPs are internalized into the cells in a similar fashion to native viruses [73]. VLPs are believed to present identical molecular recognition elements to the outside world as do native viruses.

In a recent study, the binding specificity of RHDV particles has been probed by agglutination assays [65]. It was found that RHDV recognizes A-antigen type II and H-antigen type II ABO blood group determinants. A similarity in binding specificity with UEA-I lectin is proposed based on this data. Yet, no structural similarity has been found between lectins and the P2 subunit of the protruding domain of the viral protein 60 (VP60), which covers the binding site [74]. Additionally, a strong preference of H-antigen type II over type I has been described in this study [65].

To investigate the binding specificity of the RHDV capsid at atomic resolution, we have performed preliminary STD NMR experiments that show that this system is very well suited for this type of approach [75]. Even though the application of STD NMR is not limited by increasing the size of the receptor, special care has to be taken for the setup of STD NMR experiments employing very large receptors such as viruses or VLPs. From Stokes' law, one can estimate a correlation time of a RHDV VLP in the range of 75 μs. In this motional regime, relaxation properties are dramatically different from those of "normal" molecular weight proteins, and as a consequence the conditions for saturation transfer from the receptor protein to the ligand will be different.

In general, saturation transfer is more efficient the larger the receptor. This is mainly due to a faster spread of saturation amongst protons of the receptor due to significantly enhanced spin diffusion [76, 77]. In other words, saturation is spread rather homogeneously and very rapidly over the viral shell. Therefore, subsequent intermolecular transfer of saturation onto the ligand is much less dependent on saturation transfer processes inside the receptor protein. As a consequence of the decreasing T_2 relaxation time, proton resonances of the virus are not observed due to extreme line broadening, and hence no suppression via spin locking is necessary [57]. It has to be noted that for VLPs that lack any RNA, no signals from polyamines are observed in the NMR spectrum. In the case of HRV2 such signals had interfered with the ligand signals under study [53].

Furthermore, the increased line width allows the on-resonance frequency for saturation of the virus capsid to be placed without any risk of saturating the ligand signals directly. Compared to receptors with lower tumbling times, the efficiency of saturation transfer depends much less on the position of the on-resonance frequency as this has been observed previously for STD NMR studies of the VLDLR fragment V3 with HRV2 [57]. On the other hand, the extreme line broadening requires the off-resonance frequency to be placed significantly more distant from on-resonance than usual, and it turns out that there is a significant dependence of the intensity of signals of the difference spectrum on the position of the off-resonance frequency.

As a consequence of the high symmetry of the icosahedral viral shell, there are numerous symmetry-related binding sites, leading to an increased saturation transfer to binding and exchanging ligands per particle. For the RHDV particles 90 binding sites are assumed. This is in contrast to other receptors of large molecular size with a single binding site, e.g., the ribosome [78].

Besides the benefits of the increasing correlation time of the virus, the quantitative interpretation of the saturation transfer might be limited by the breakdown of the traditional relaxation theory based on the "motional nar-

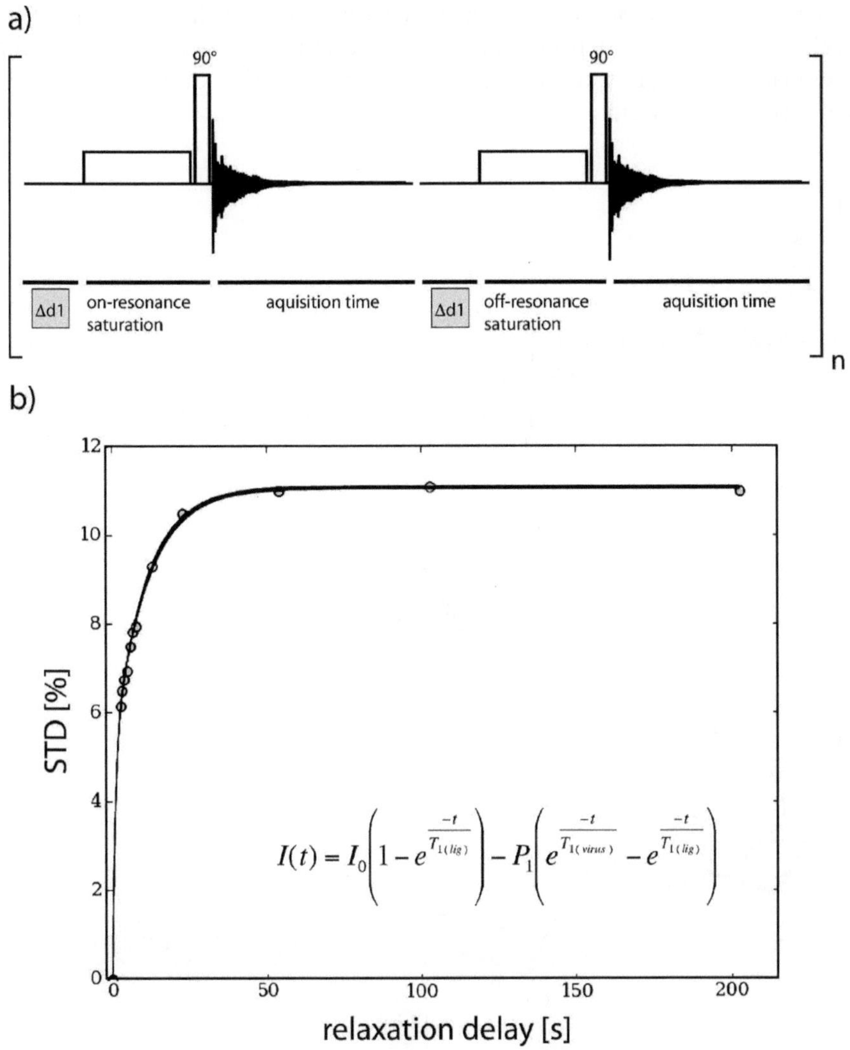

Fig. 5 Effect of varying relaxation delays between on- and off-resonance experiments in STD NMR experiments. **a** Experimental setup for interleaved measurements in STD NMR spectroscopy. n represents the number of scans. The inter-scan delay $\Delta d1$ is varied while keeping on- and off-resonance frequencies constant at –4 and +300 ppm, respectively. **b** The resulting STD effects for the *O*-methyl group of α-L-Fuc-*O*-methyl in the presence of RHDV VLPs. The equation that was used for non-linear least squares data fitting is based on the saturation recovery experiment [98]. With $T_{1(\mathrm{lig})} = 0.91$ s as measured independently (unpublished results) and a Monte Carlo error estimation yields $T_{1(\mathrm{virus})} = 10.06 \pm 0.41$ s. This value does not directly correspond to a T1 relaxation time of the virus protons, because other factors also influence the observed relaxation [99]. According to these findings a relaxation delay $\Delta d1 = 25$ s was employed in all STD experiments. This results in a recovery of 92% of the virus resonance, and thereby reduces errors in epitope mapping that are introduced otherwise by non-homogeneous recovery of the binding site.

rowing approximation" [79]. With the current theory at hand, it cannot be excluded that an r^{-6} dependency of the saturation transfer is not valid for these large receptor systems. A qualitative interpretation of the binding epitopes certainly is valid.

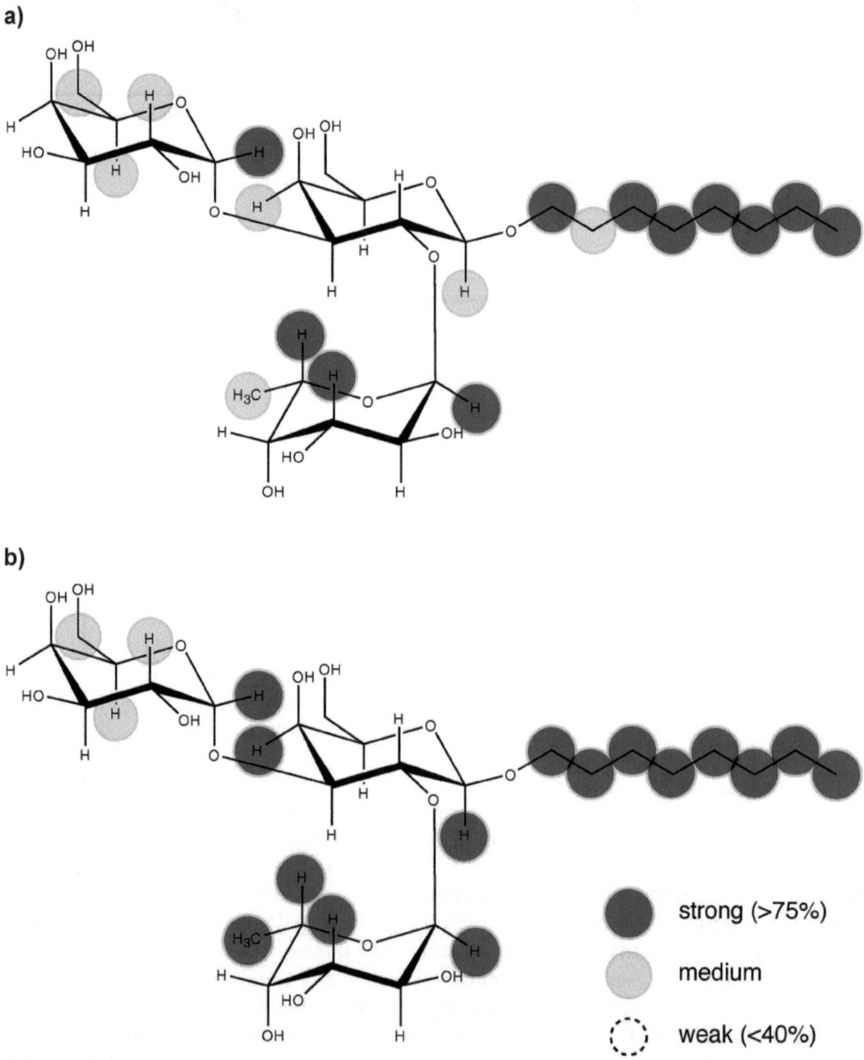

Fig. 6 Binding epitopes for the B-antigen trisaccharide in the presence of RHDV VLPs recorded with **a** $\Delta d1 = 20\ \mu s$ and **b** $\Delta d1 = 100$ s. A setting of $\Delta d1 = 20\ \mu s$ for the interpulse relaxation delay corresponds to a total relaxation delay $t = 4.7$ s since relaxation also occurs during the acquisition time. With $\Delta d1 = 100$ s an increase of STD effects for H2, H1′, H4′ and H6‴ is observed. This is due to reduced saturation during the off-resonance experiment resulting from a sufficient long relaxation delay

Another parameter of critical importance is the T_1 relaxation time. T_1 relaxation times increase with increasing correlation times. Therefore, long relaxation delays are required for large receptor proteins when STD spectra are recorded in an interleaved manner, alternating between on- and off-resonance experiments (Fig. 5a). Otherwise unwanted saturation will partially persist in the off-resonance experiment. Experimentally, we observe differences in STD intensities until delays of about 25 s are employed (Fig. 5b). It cannot be expected that a homogeneous T_1 relaxation time is found for all receptor protons in the binding pocket. From theoretical calculations Ishima and coworkers deduced that methyl groups act as relaxation sinks and should therefore lower the T_1 relaxation time to approximately 1.25 s [80]. Therefore, the saturation present during the off-resonance experiment is assumed to affect the binding epitope non-homogeneously if the recycling delay is set at inappropriate values (Fig. 6a,b).

Following these experimental guidelines it will be possible to analyze the binding epitopes of HBGA ligands of RHDV at atomic resolution.

2.4 Noroviruses

Another genus of the *Caliciviridae* family is represented by the noroviruses. Using STD NMR approaches it should be possible to tackle some of the open questions concerning ligand recognition and discrimination by noroviruses [64, 81, 82]. Such experiments are underway. Here we will only give a short overview of the open questions and potential applications of such investigations.

The different genotypes display a high mutation rate, turning them into very adaptable pathogens in the humane intestine. Therefore, it is not surprising that noroviruses are the major cause of gastroenteritis. The occurrence of novel variants, so-called epidemic variants, means a constant challenge for the human body in evading noroviral infection [83]. Furthermore, it is suggested that there is an animal reservoir of noroviruses capable of crossing the species barrier [64, 84]. Additionally, there is also a human reservoir maintaining a basal level of norovirus infections in a population [85, 86]. The combination of these three factors results in a high susceptibility to noroviruses, which leads to outbreaks in hospitals, schools, cruise ships and military installations [87].

To overcome the burden of host diversity, noroviruses have developed strategies that turn these viruses into a source of common and very widespread disease [60, 82]. Although cluster analysis of recombinant VLPs improves our understanding of the correlation between genetic diversity and molecular recognition of different ligands [88], a profound analysis of the host/pathogen interplay on a molecular level is required. An 8% sequence identity in the carbohydrate-binding domain between noroviruses renders this kind of analysis strikingly difficult [81].

Ligand-based NMR techniques are well suited for investigation of the carbohydrate binding process. The presence of two distinct binding sites has been suggested, based on data from oligosaccharide-based assays [89]. In this model, the presence of a strain-specific binding site in conjunction with a binding site for H-antigen is proposed [64, 89]. This can be investigated by STD NMR titration experiments as shown before for HRV2 [53]. Noroviruses with a preference for Lewis antigens show inhibition of binding in the presence of the A/B blood group determinants [81]. This so-called epitope-masking effect was observed in saliva assays [81] and can also be subjected to scrutiny by STD NMR experiments.

Another long-standing question concerns the influence of residues other than the terminal HBGAs [74]. It has been observed that HBGAs attached to BSA and polyphenylacetylene have higher affinities as free oligosaccharides [90]. An analysis covering extended ligands will provide valuable information on their recognition and on the importance of larger oligosaccharide moieties. A recent study has presented first leads for rational drug design [91]. Various NMR screening approaches are possible for identifying binding HBGAs or to identify small organic compounds as hits for fragment-based entry inhibitor design process [92]. Moreover, the identification of binding carbohydrates from milk oligosaccharide mixtures, which are known to protect against norovirus infection, would be feasible by this approach [63].

3 Conclusion

Studying the molecular recognition of ligands by native viruses and VLPs offers structural insight into the viral attachment and entry processes. Details at atomic resolution are important for our understanding of these early events during the viral life cycle [6]. Drug development benefits from this effort in order to develop entry inhibitors.

The large molecular size of viruses compared to traditional drug targets hampers the routine application of X-ray crystallography and receptor-based NMR techniques. On the other hand, it had been demonstrated that ligand-based NMR experiments are a powerful tool for studying the binding of ligands to native viruses [53]. Especially, for non-enveloped viruses such as picornaviruses or caliciviruses such approaches will be very powerful.

For enveloped viruses such as the influenza viruses it has been shown that similar binding affinities are found for the hemagglutinin (HA) ligand sialic acid using either isolated HA or the whole virus [37, 40] using simple ^{1}H NMR titration experiments. In contrast to structural proteins of non-enveloped viruses, HA is a membrane protein and thus not as rigid as proteins as part of, e.g., icosahedral particles. For the application of STD NMR experiments this may circumvent the peculiarities such as very long relaxation delays that are required because of long rotational diffusion times of non-enveloped viruses.

STD NMR is based on the accumulation of saturated ligands in solution. Hence, working with very low receptor quantities is one of the great benefits of this technique. Therefore, especially for systems where the amount of material is a limiting factor, such as viruses or VLPs, these experiments will be very helpful.

The ligand-based NMR studies on virus–ligand interactions published so far indicate that experimental conditions have to be chosen carefully, as described above. Once proper conditions are met, STD NMR lends itself especially well to systematic studies of ligand binding to viruses and VLPs. We are convinced that in the future the combination of ligand-based NMR experiments and X-ray crystallography will improve our understanding of viral entry mechanisms.

Acknowledgements CR thanks the Fonds der chemischen Industrie for a stipend. TP thanks the Deutsche Forschungsgemeinschaft (Grants Me 1830/1 and HBFG 101/192-1) and the University of Luebeck for support.

References

1. Dimitrov DS (2004) Nat Rev Microbiol 2:109
2. Sieczkarski SB, Whittaker GR (2005) Curr Top Microbiol Immunol 285:1
3. Stewart PL, Dermody TS, Nemerow GR (2003) Adv Protein Chem 64:455
4. Baranowski E, Ruiz-Jarabo CM, Pariente N, Verdaguer N, Domingo E (2003) Adv Virus Res 62:19
5. Marsh M, Helenius A (2006) Cell 124:729
6. Damm EM, Pelkmans L (2006) Cell Microbiol 8:1219
7. Gagneux P, Cheriyan M, Hurtado-Ziola N, van der Linden EC, Anderson D, McClure H, Varki A, Varki NM (2003) J Biol Chem 278:48245
8. Baranowski E, Ruiz-Jarabo CM, Domingo E (2001) Science 292:1102
9. Flint SJ (2004) Principles of virology: molecular biology, pathogenesis, and control of animal viruses. ASM, Washington DC
10. Hogle JM (2002) Annu Rev Microbiol 56:677
11. Dormitzer PR, Nason EB, Prasad BV, Harrison SC (2004) Nature 430:1053
12. Vliegenthart GA, Gompper G (2006) Biophys J 91:834
13. Chow M, Basavappa R, Hogle JM (1997) The role of conformational transitions in poliovirus pathogenesis. In: Chiu W, Garcea R, Burnette R (eds) Structural biology of viruses. Oxford University Press, Oxford, pp 157–186
14. Fox MP, Otto MJ, McKinlay MA (1986) Antimicrob Agents Chemother 30:110
15. Kim SS, Smith TJ, Chapman MS, Rossmann MC, Pevear DC, Dutko FJ, Felock PJ, Diana GD, McKinlay MA (1989) J Mol Biol 210:91
16. Smith TJ, Kremer MJ, Luo M, Vriend G, Arnold E, Kamer G, Rossmann MG, McKinlay MA, Diana GD, Otto MJ (1986) Science 233:1286
17. Tsang SK, Danthi P, Chow M, Hogle JM (2000) J Mol Biol 296:335
18. Phelps DK, Rossky PJ, Post CB (1998) J Mol Biol 276:331

19. Munch J, Standker L, Adermann K, Schulz A, Schindler M, Chinnadurai R, Pohlmann S, Chaipan C, Biet T, Peters T, Meyer B, Wilhelm D, Lu H, Jing W, Jiang S, Forssmann WG, Kirchhoff F (2007) Cell 129:263
20. O'Hara BM, Olson WC (2002) Curr Opin Pharmacol 2:523
21. Welch BD, Vandemark AP, Heroux A, Hill CP, Kay MS (2007) Proc Natl Acad Sci USA 104:16828
22. Bomsel M, Alfsen A (2003) Nat Rev Mol Cell Biol 4:57
23. The Scripps Research Institute (2007) Viperdb:Virus particle explorer. TSRI, La Jolla, CA. http://viperdb.scripps.edu. Last visited: 12 Dec 2007
24. Natarajan P, Lander GC, Shepherd CM, Reddy VS, Brooks CL III, Johnson JE (2005) Nat Rev Microbiol 3:809
25. Wall ME, Gallagher SC, Trewhella J (2000) Annu Rev Phys Chem 51:355
26. Rossmann MG, Morais MC, Leiman PG, Zhang W (2005) Structure 13:355
27. Meinhold L, Smith JC (2006) Proteins, p 941
28. Verdaguer N, Fita I, Reithmayer M, Moser R, Blaas D (2004) Nat Struct Mol Biol 11:429
29. Fry EE, Lea SM, Jackson T, Newman JW, Ellard FM, Blakemore WE, Abu-Ghazaleh R, Samuel A, King AM, Stuart DI (1999) EMBO J 18:543
30. Xiao C, Bator-Kelly CM, Rieder E, Chipman PR, Craig A, Kuhn RJ, Wimmer E, Rossmann MG (2005) Structure 13:1019
31. Szymczyna BR, Gan L, Johnson JE, Williamson JR (2007) J Am Chem Soc 129:7867
32. Doherty PC, Turner SJ, Webby RG, Thomas PG (2006) Nat Immunol 7:449
33. Palese P (2004) Nat Med 10:S82
34. Matrosovich M, Tuzikov A, Bovin N, Gambaryan A, Klimov A, Castrucci MR, Donatelli I, Kawaoka Y (2000) J Virol 74:8502
35. Vines A, Wells K, Matrosovich M, Castrucci MR, Ito T, Kawaoka Y (1998) J Virol 72:7626
36. Angata T, Varki A (2002) Chem Rev 102:439
37. Hanson JE, Sauter NK, Skehel JJ, Wiley DC (1992) Virology 189:525
38. von Itzstein M, Wu WY, Kok GB, Pegg MS, Dyason JC, Jin B, Van Phan T, Smythe ML, White HF, Oliver SW et al. (1993) Nature 363:418
39. Russell RJ, Haire LF, Stevens DJ, Collins PJ, Lin YP, Blackburn GM, Hay AJ, Gamblin SJ, Skehel JJ (2006) Nature 443:45
40. Sauter NK, Bednarski MD, Wurzburg BA, Hanson JE, Whitesides GM, Skehel JJ, Wiley DC (1989) Biochemistry 28:8388
41. Reibarkh M, Malia TJ, Wagner G (2006) J Am Chem Soc 128:2160
42. Semler BL, Wimmer E (2002) Molecular biology of picornaviruses. ASM, Washington DC
43. Tan WC (2005) Curr Opin Pulm Med 11:21
44. Whitton JL, Cornell CT, Feuer R (2005) Nat Rev Microbiol 3:765
45. Modrow S, Falke D, Truyen U (2003) Molekulare Virologie. Spektrum Akad, Heidelberg
46. Staunton DE, Gaur A, Chan PY, Springer TA (1992) J Immunol 148:3271
47. Andries K, Dewindt B, De Brabander M, Stokbroekx R, Janssen PA (1988) Arch Virol 101:155
48. Goncalves RB, Mendes YS, Soares MR, Katpally U, Smith TJ, Silva JL, Oliveira AC (2007) J Mol Biol 366:295
49. Lewis JK, Bothner B, Smith TJ, Siuzdak G (1998) Proc Natl Acad Sci USA 95:6774
50. Goncalves RB, Mendes YS, Soares MR, Katpally U, Smith TJ, Silva JL, Oliveira AC (2007) J Mol Biol 366:295

51. Korant BD, Lonberg-Holm K, Noble J, Stasny JT (1972) Virology 48:71
52. Lonberg-Holm K, Noble-Harvey J (1973) J Virol 12:819
53. Benie AJ, Moser R, Baeuml E, Blaas D, Peters T (2003) J Am Chem Soc 125:14
54. Bella J, Kolatkar PR, Marlor CW, Greve JM, Rossmann MG (1998) Proc Natl Acad Sci USA 95:4140
55. Yamaya M, Sasaki H (2003) Viral Immunol 16:99
56. Hewat EA, Neumann E, Conway JF, Moser R, Ronacher B, Marlovits TC, Blaas D (2000) EMBO J 19:6317
57. Moser R, Snyers L, Wruss J, Angulo J, Peters H, Peters T, Blaas D (2005) Virology 338:259
58. Blacklow SC (2004) Nat Struct Mol Biol 11:388
59. Jeon H, Blacklow SC (2005) Annu Rev Biochem 74:535
60. Marionneau S, Cailleau-Thomas A, Rocher J, Le Moullac-Vaidye B, Ruvoen N, Clement M, Le Pendu J (2001) Biochimie 83:565
61. Le Pendu J, Marionneau S, Cailleau-Thomas A, Rocher J, Le Moullac-Vaidye B, Clement M (2001) Apmis 109:9
62. Hakomori S (1999) Biochim Biophys Acta 1473:247
63. Le Pendu J (2004) Adv Exp Med Biol 554:135
64. Tan M, Jiang X (2005) Trends Microbiol 13:285
65. Ruvoen-Clouet N, Ganiere JP, Andre-Fontaine G, Blanchard D, Le Pendu J (2000) J Virol 74:11950
66. Moss SR, Turner SL, Trout RC, White PJ, Hudson PJ, Desai A, Armesto M, Forrester NL, Gould EA (2002) J Gen Virol 83:2461
67. Capucci L, Fusi P, Lavazza A, Pacciarini ML, Rossi C (1996) J Virol 70:8614
68. Thiel HJ, Konig M (1999) Vet Microbiol 69:55
69. Tan M, Huang P, Meller J, Zhong W, Farkas T, Jiang X (2003) J Virol 77:12562
70. Jiang X, Wang M, Graham DY, Estes MK (1992) J Virol 66:6527
71. Jiang X, Matson DO, Ruiz-Palacios GM, Hu J, Treanor J, Pickering LK (1995) J Clin Microbiol 33:1452
72. Jiang X, Zhong WM, Farkas T, Huang PW, Wilton N, Barrett E, Fulton D, Morrow R, Matson DO (2002) Arch Virol 147:119
73. Ewers H, Smith AE, Sbalzarini IF, Lilie H, Koumoutsakos P, Helenius A (2005) Proc Natl Acad Sci USA 102:15110
74. Cao S, Lou Z, Tan M, Chen Y, Liu Y, Zhang Z, Zhang XC, Jiang X, Li X, Rao Z (2007) J Virol 81:5949
75. Angulo J, Rademacher C, Biet T, Benie AJ, Blume A, Peters H, Palcic M, Parra F, Peters T (2006) Methods Enzymol 416:12
76. Kalk A, Berendsen HJC (1976) J Magn Reson 24:343
77. Andree PJ (1978) J Magn Reson 29:419
78. Peng JW, Moore J, Abdul-Manan N (2004) Prog Nucl Magn Reson Spectr 44:225
79. Slichter CP (1996) Principles of magnetic resonance. Springer, Berlin, Heidelberg, New York
80. Ishima R, Shibata S, Akasaka K (1991) J Magn Reson 91:455
81. Tan M, Jiang X (2007) Expert Rev Mol Med 9:1
82. Estes MK, Prasad BV, Atmar RL (2006) Curr Opin Infect Dis 19:467
83. Siebenga JJ, Vennema H, Renckens B, de Bruin E, van der Veer B, Siezen RJ, Koopmans M (2007) J Virol 81:9932
84. Farkas T, Nakajima S, Sugieda M, Deng X, Zhong W, Jiang X (2005) J Clin Microbiol 43:657
85. Gallimore CI, Lewis D, Taylor C, Cant A, Gennery A, Gray JJ (2004) J Clin Virol 30:196

86. Nilsson M, Hedlund KO, Thorhagen M, Larson G, Johansen K, Ekspong A, Svensson L (2003) J Virol 77:13117
87. Lindesmith L, Moe C, Marionneau S, Ruvoen N, Jiang X, Lindblad L, Stewart P, Le Pendu J, Baric R (2003) Nat Med 9:548
88. Chakravarty S, Hutson AM, Estes MK, Prasad BV (2005) J Virol 79:554
89. Huang P, Farkas T, Zhong W, Tan M, Thornton S, Morrow AL, Jiang X (2005) J Virol 79:6714
90. Tan M, Hegde RS, Jiang X (2004) J Virol 78:6233
91. Feng X, Jiang X (2007) Antimicrob Agents Chemother 51:324
92. Hajduk PJ, Greer J (2007) Nat Rev Drug Discov 6:211
93. Verdaguer N, Blaas D, Fita I (2000) J Mol Biol 300:1179
94. Stierand K, Rarey M (2007) Chem Med Chem 2:853
95. Rarey M, Kramer B, Lengauer T, Klebe G (1996) J Mol Biol 261:470
96. Wang R, Lai L, Wang S (2002) J Comput Aided Mol Des 16:11
97. Mayer M, Meyer B (2001) J Am Chem Soc 123:6108
98. Freeman R, Hill HDW (1971) J Chem Phys 54:3367
99. Jayalakshmi V, Krishna NR (2002) J Magn Reson 155:106

Top Curr Chem (2008) 273: 203–214
DOI 10.1007/128_2007_21

Published online: 22 February 2008

Investigation of Proteins in Living Bacteria with In-Cell NMR Experiments

Volker Dötsch

Institute of Biophysical Chemistry, University of Frankfurt, Max-von-Laue-Str. 9, 60438 Frankfurt, Germany
vdoetsch@em.uni-frankfurt.de

Abstract In recent years NMR methods have been developed that enable the observation of proteins inside living bacterial cells. Because of the sensitivity of the chemical shift to environmental changes these in-cell NMR experiments can be used to study protein conformation, molecular interaction or dynamics in a protein's natural surrounding. Detection of proteins in the bacterial cytoplasm relies on labeling of the protein of interest with NMR active isotopes. This review describes different labeling techniques based on either uniform ^{15}N or ^{13}C labeling as well as amino acid specific labeling schemes. In addition potential applications of these in-cell NMR experiments and their limitations are discussed.

Keywords Drug interaction · In-cell NMR · Isotope labeling · Protein conformation

1 Introduction

During the last three decades NMR spectroscopy has developed into a very important analytical tool both in the chemical and biological sciences. The most important aspect is its ability to determine structures of biological macromolecules under near physiological conditions [1]. In addition, NMR spectroscopy is capable of investigating the interaction between biological macromolecules and other molecules, ranging from proteins and nucleic acids to small organic compounds. The ability to investigate inter-molecular interactions is based on the sensitivity of the chemical shift to changes in the chemical environment [2–5]. In addition to direct interaction with other molecules

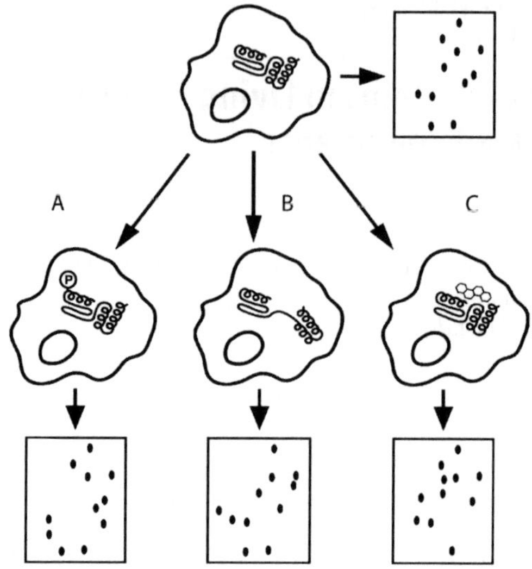

Fig. 1 Potential applications for in-cell NMR experiments. Changes in the chemical environment of a protein's nuclei caused for example by post-translational modifications (**A**), conformational changes (**B**) or binding events (**C**) can be detected by differences in chemical shifts in in-cell NMR experiments. Schematic HSQC spectra indicating the sensitivity of the chemical shift to the changes described above are shown next to each cell. Reprinted with permission from Serber and Dötsch, 2001

changes of the chemical shifts of a biological macromolecule can also indicate changes of its conformation, for example different folding states (Fig. 1). So far, most binding studies using NMR spectroscopy have been carried out in in vitro experiments. Investigating molecular interactions directly in living cells relies mainly on either indirect methods such as colocalization studies of fluorescently labeled molecules or on immunoprecipitation assays, which are, however, strictly speaking not interaction studies in living cells. The recent development of in-cell NMR experiments that enable NMR investigations of biological macromolecules in their natural environment has created new tools for the direct investigation of the interaction between biological macromolecules and other macromolecules or drugs in living cells [6–20]. In addition, these methods allow researchers to obtain information about the conformation and the dynamics of macromolecules in their natural environment and to compare the in vivo behavior with data obtained from in vitro studies.

2
Basic Concept

The observation of biological macromolecules in cellular systems is based on labeling schemes that enable the selective identification of these macromolecules in an environment that is crowded with other macromolecules as well as with many different small molecules [6, 10, 11]. In principle, distinguishing the NMR resonances of the molecule of interest from all other molecules can be achieved in two different ways, both of which are based on labeling the protein of interest with NMR-active isotopes, in particular ^{15}N and ^{13}C. The first one is based on micro-injecting a highly concentrated sample of a purified macromolecule labeled with NMR active isotopes into the cell type of interest [21, 22]. In addition to direct injection, the use of arginine-tags or dendrite-based systems might also be possible for bringing bacterially expressed and isotopically labeled proteins into cells. The advantage of these techniques is that the natural abundance of the NMR active isotopes (1.1% ^{13}C and 0.4% ^{15}N) represents the only potential background signals. However, these techniques are very labor intensive and technically very demanding and are therefore just emerging. The method that has been used most often up to now is the expression of the macromolecule of interest inside cells, mainly in bacterial cells [6–20]. While this method is convenient and uses very well-established expression protocols it has the potential of producing a huge amount of background signals. Interestingly, experiments with uniform ^{15}N-labeling of the over-expressed protein have demonstrated that despite the fact that the entire cell with all its components is labeled with ^{15}N only a minimal level of background signals is produced [7] (Fig. 2). The only requirements for observing a protein in amide proton-detected in-cell NMR experiments is that its expression level reaches a certain threshold and that its rotational correlation time is sufficiently short. The last condition is met for all macromolecules that do not interact with large cellular components since the intracellular viscosity is only a factor of two higher than the viscosity of water [14].

In contrast, labeling with ^{13}C results in a high level of background signals that makes an unambiguous identification of the signals of the protein of interest impossible (Fig. 3). One exception are signals with unique chemical shifts such as high-field shifted methyl groups or anomeric sugar resonances [9, 13]. This surprising difference in the level of background signals between ^{15}N and ^{13}C is most likely based on the fact that C–H groups are far more abundant in small molecules in the cell than N–H groups. In addition, amide protons that are not protected from solvent exchange through involvement in hydrogen bonds exchange fast with the bulk water, thus making their NMR signals undetectable due to exchange line broadening.

Fortunately, the problem of high background signals can be overcome by amino acid-type selective labeling schemes [7, 13]. However, the type of amino acids chosen for the selective labeling has to be considered carefully since some amino acids are used as metabolic precursors for other amino acid types thus resulting in labeling more than one kind of amino acid. Amino acids that are at the end of a metabolic pathway and can thus be used for in-cell NMR experiments are Lys, Arg or His for ^{15}N labeling and Met

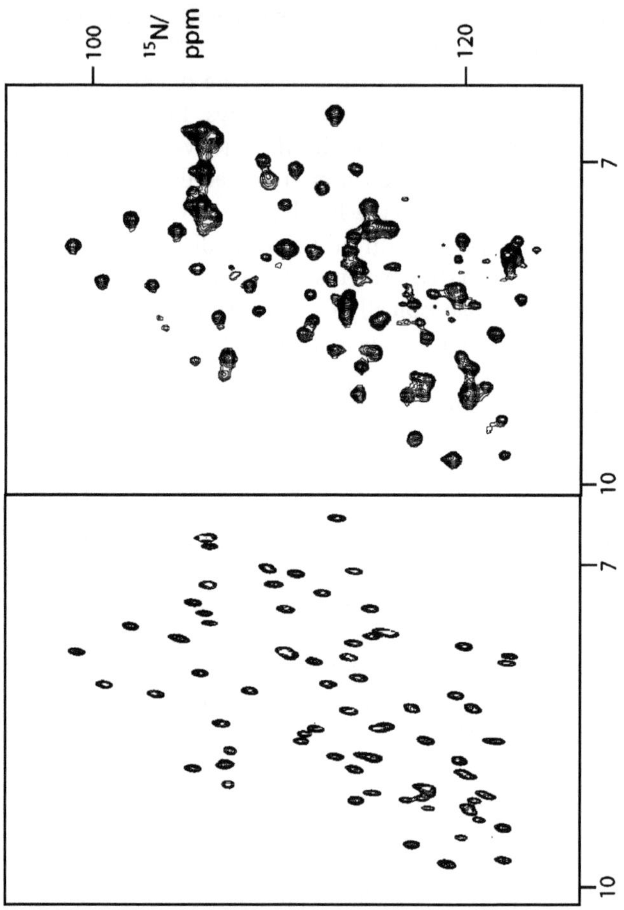

Fig. 2 Comparison of an HSQC spectrum of a purified in vitro sample of NmerA (*left*) with an in-cell spectrum of NmerA (*right*). The level of background signals in the in-cell spectrum is low, allowing the observation of the protein in its natural environment

and with some restriction Ala for ^{13}C labeling (Fig. 4). Other amino acids can be produced selectively from labeled precursors and are therefore good candidates for in-cell NMR experiments. One example is the conversion of α-ketobutyrate to isoleucine by bacteria [23]. In case other amino acid types should be labeled auxotrophic strains that are incapable of producing certain types of amino acids can be used [24, 25]. However, the expression yield of such strains is usually reduced, leading to a reduced sensitivity of the in-cell NMR experiments.

In addition to using auxotrophic strains of naturally occurring amino acids, the laboratory of Peter Schultz has developed very elegant methods to incorporate labeled amino acids in specific positions of a protein by extending the genetic code of prokaryotic as well as eukaryotic cells [26, 27]. The

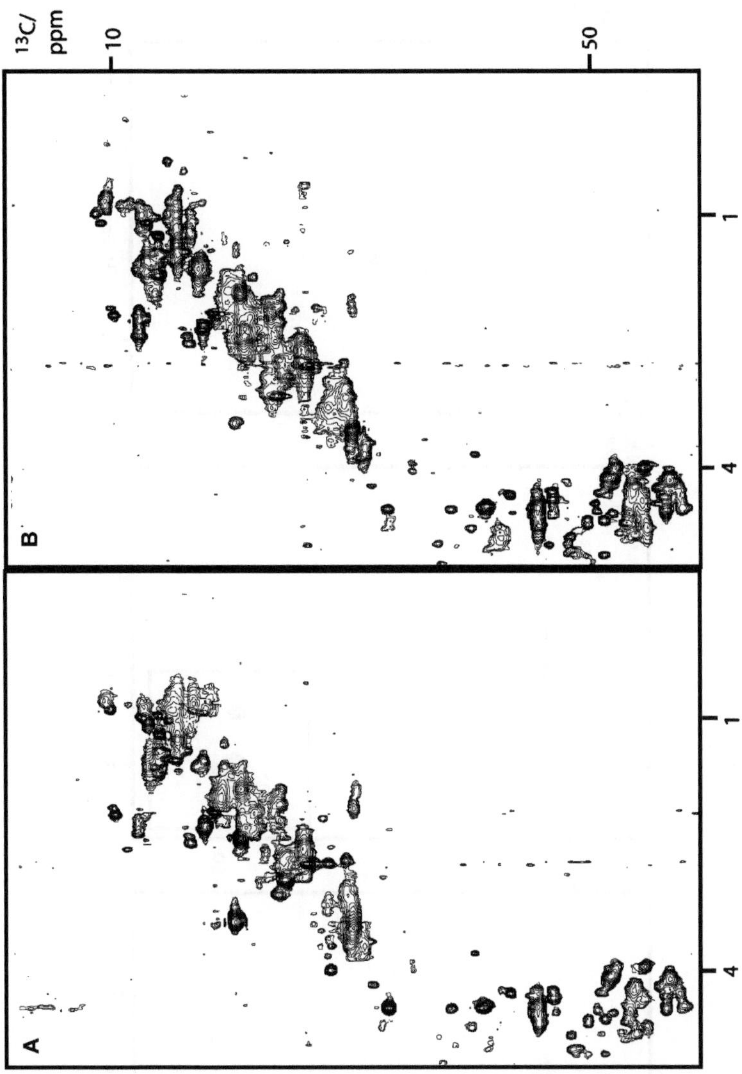

Fig. 3 Sections of two [^{13}C,^{1}H]-HSQC spectra taken from a bacterial sample grown on minimal medium containing ^{13}C-labeled glucose with (**A**) and without (**B**) over-expression of calmodulin. Reprinted with permission from Serber et al., 2004

disadvantage, however, is the requirement for using non-natural amino acids that can also change the behavior of a protein [27].

Instead of using amino acid-type selective labeling suppression of high background signals can also be achieved by selective expression of the protein of interest with the concomitant suppression of the expression of all other (host-) proteins. In principle, this suppression of host gene expres-

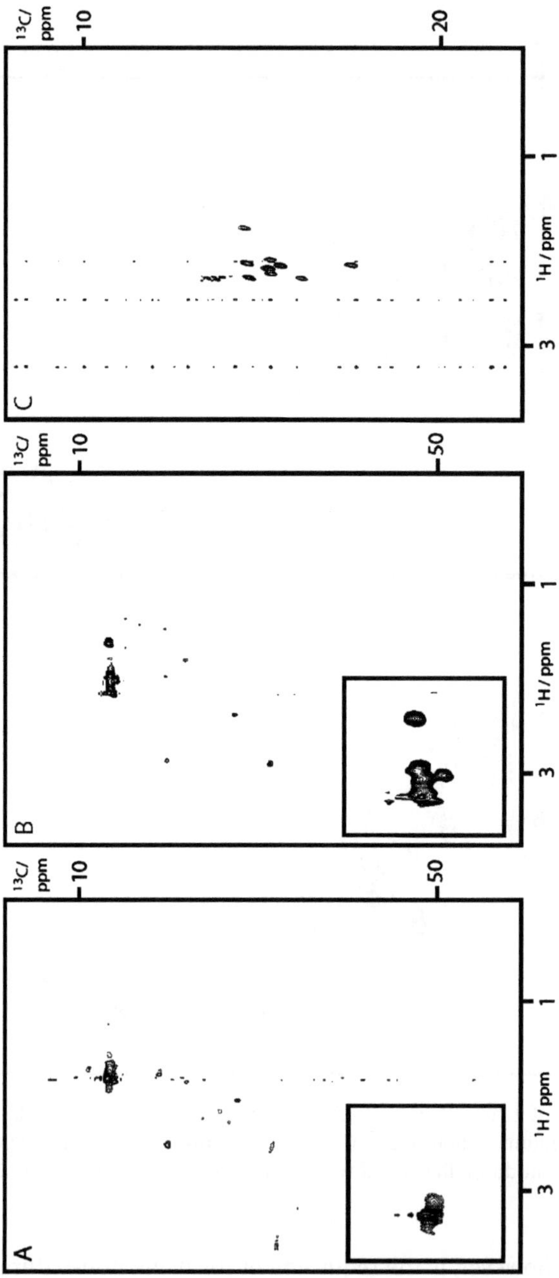

Fig. 4 Comparison of background levels in [^{13}C,^{1}H]-HSQC spectra of bacterial samples grown on ^{13}C-methyl group labeled methionine without over-expression (**A**) and with over-expression of calmodulin (**B**). The spectrum in **C** was measured with a smaller spectral width as the spectrum in **B**. Only the free methionine and the nine methionine resonances of calmodulin are visible. Reprinted with permission from Serber et al., 2004

sion is possible with two different techniques. The first is based on using the drug rifampicin which inhibits the bacterial RNA-polymerase [7, 28, 29]. The RNA polymerase of the bacteriophage T7, however, is not affected by this drug, thus allowing the selective over-expression of the protein of interest from a plasmid containing a T7 promoter. The second approach that has very recently been described is the degradation of all mRNA that contains ACA sequences by an mRNA Interferase [30]. Changing all ACA base triplets in the DNA sequence of the protein of interest by utilizing other codons enables the selective translation of the mRNA of a particular protein while all other mRNAs that contain ACA triplets are degraded.

3
Protein Conformations

The goal of "in-cell NMR" experiments is not to determine structures directly in the cellular environment, but to use the sensitivity of the chemical shift towards changes in the environment to investigate the behavior of a macromolecule in its natural surroundings. Any changes of the conformation, post-translational modification or binding event results in changes of the chemical shifts. In case differences between the in-vitro spectra and the in-cell spectra can be detected, further in vitro experiments are necessary to identify the cause of these differences [7, 10, 19]. In these in vitro experiments the suspected interaction partners are added to the NMR sample to see if the same chemical shift differences are produced. Using this strategy, Dedmon et al. could show that the bacterial protein FlgM which is completely unfolded in vitro is partially folded in the *E. coli* cytoplasm [10]. They could reproduce the spectral characteristics of the in-cell sample by adding high concentrations of either other proteins (BSA) or small molecules (sugar) to an in vitro sample of FlgM. On the basis of this behavior they concluded that the high concentration of other (macro-) molecules inside the cell, known as molecular crowding, is responsible for the observed partial folding. Molecular crowding, however, does not always lead to a more compact structure of natively disordered proteins. Detailed investigation of the in vitro and in vivo behavior of α-synuclein demonstrated that this protein maintains its disordered conformation at all temperatures when expressed in the *E. coli* periplasm while it shows a temperature-induced formation of secondary structure elements in vitro [19].

If substantial conformational differences between the in vivo and the in vitro states of the protein exist the in vitro assignments of the chemical shifts might not be transferable to the in-cell NMR spectra, making a new chemical shift assignment necessary. One of the main difficulties of measuring multi-dimensional NMR spectra of proteins inside living cells is that the life time of the cells or of the protein inside the cells is often smaller than the measurement time for the average multi-dimensional NMR experiment.

Recently, however, NMR techniques have been developed that can be used to considerably accelerate the measurement of large three-dimensional data sets [31, 32]. These reduced-dimensionality techniques have already been successfully used in in-cell NMR applications [16]. In the future these techniques might enable the complete backbone assignment of a protein in its natural environment and might enable detailed investigation of different folding states.

4
Protein–Protein Interaction

As discussed in the introduction differences in chemical shifts can be caused by conformational changes as well as binding events. Understanding the biological function of a protein requires a detailed knowledge of all interactions of a given protein with the other cellular components. In-cell NMR spectroscopy can—in principle—be used to investigate this interaction with other cellular components with the limitation that the resultant complexes must tumble sufficiently fast to be observable by high-resolution liquid-state NMR spectroscopy. One disadvantage of NMR spectroscopy is its inherently low sensitivity which requires concentrations at least in the tens of μM range. As discussed above, in-cell NMR spectroscopy is based on the over-expression of proteins, bringing them to the required concentration. If protein–protein interactions are to be studied by in-cell NMR spectroscopy, the interaction partner has to reach similar concentrations. The research group of Alexander Shekthman has achieved this by using a sequential expression system [20]. After expressing the first protein in a labeled medium, changing the medium and inducing the over-expression of the interacting protein in a non-labeled medium allows the interaction between both proteins to be monitored by recording in-cell NMR spectra. Changes in the chemical shifts of the first, labeled protein indicate an interaction. Using this method with a L-arabinose inducible and an IPTG-inducible promoter allowed Alexander Shekthman and co-workers to investigate the interaction between ubiquitin and two interaction partners, STAM2 and a peptide derived from the ataxin 3 protein in bacterial cells.

In addition to these specific interactions, many proteins show non-specific interactions with their cellular surroundings. Examples are FKBP and thioredoxin which cannot be observed by amide proton-based NMR spectroscopy because they interact with other cellular components [13]. Labeling of methyl groups with ^{13}C, however, makes certain amino acids (such as methionine and isoleucine) observable in living cells since the inherent fast rotation of the methyl group enables detection of methyl group resonances even of very large complexes [33–35]. In addition, the sensitivity of methyl groups relative to amide protons is further increased due to the higher number of protons directly attached to the hetero-atom and the fact that the sensitivity of amide proton detection is often reduced due to chemical exchange with water. Methyl groups, in particular the methyl group of methionines as well as the δ-methyl group of isoleucines show very favorable relaxation behavior, making them ideal probes for the investigation of proteins that form larger complexes in living cells.

5 Protein Drug Screens

NMR spectroscopy has become a standard tool in the pharmaceutical industry for the investigation of protein–drug interactions [3, 4, 35]. These interaction studies are, however, performed in vitro. The main disadvantage is that interactions that are observed in these experiments might not occur in vivo. A drug might for example not be able to pass the cellular membrane, it might be pumped out of the cell very fast or it might get metabolized or interact with other cellular components more efficiently than with the intended target protein. These disadvantages could be overcome if protein–drug interactions could be studied directly inside living cells by in-cell NMR spectroscopy. Hubbard et al. have reported an example of such an in-cell NMR screening application [11]. They could show that the drug BRL-16492PA that binds to the bacterial two-component signal transduction protein CheY in vitro also binds to the same protein inside living *E. coli* bacteria. They based their conclusion on observing virtually identical chemical shift changes in the [^{15}N,^{1}H]-HSQC spectrum of CheY upon adding the drug either to a purified in vitro sample or to a slurry of *E. coli* over-expressing the protein.

Most chemical shift mapping applications such as protein–drug screens have so far relied on [^{15}N,^{1}H]-HSQC experiments due to the excellent chemical shift dispersion of the amide protons and nitrogens. As an alternative methyl group-based NMR experiments have been employed in the study of protein–drug interactions due to the high sensitivity and involvement of methyl groups in drug binding. The use of methyl groups as indicators for binding is further supported by an investigation that has demonstrated that within a set of 191 crystal structures of protein–ligand complexes 92% of the ligands had a heavy atom within 6 Å of a methyl group carbon while only 82% had a heavy atom in the same distance of a backbone nitrogen [35]. To test the applicability of methyl group-based [^{13}C,^{1}H]-HSQC experiments in in-cell NMR drug screens we have investigated the interaction of calmodulin with the known drug phenoxybenzamine hydrochloride which is assumed to bind to a hydrophobic pocket that is lined with methionines [13]. We added the drug to an *E. coli* culture expressing calmodulin half an hour prior to sample preparation. Although no differences in chemical shifts between an in vitro sample and the in-cell sample could be detected, some of the peaks in the in-cell spectrum showed increased line broadening, suggesting that a weak interaction with the drug exists. Further investigations have demonstrated that phenoxybenzamine hydrochloride is mainly associated with the bacterial membrane and that the high local concentration of phenoxybenzamine near the bacterial membrane is most likely responsible for the observed weak interaction.

In addition to protein–drug interactions, in-cell NMR experiments can also be used to monitor the interaction between proteins and metal ions in living cells. One example is presented by Hubbard and coworkers who have investigated the ion binding status of the bacterial two-component signal transduction protein CheY in the bacterial cytoplasm [11]. By comparing an in-cell [^{15}N,^{1}H]-HSQC spectrum with in vitro spectra of the protein complexed with different metal ions they could demonstrate that CheY preferentially binds Mg^{2+} ions in the *E. coli* cytoplasm. They also observed small additional changes in the chemical shifts that might indicate further interactions with other components. These additional shifts, however, could not be conclusively interpreted so far. Our own investigations of the calcium-binding protein calmodulin in living *E. coli* had indicated that calmodulin mainly exists in the apo-form, showing that the intracellular Ca^{2+} concentration in bacteria is not high enough to significantly populate the metal ion-bound form of the protein [7]. However, we also observed additional peaks that indicate that more than one conformation is present under these conditions in the bacterial cytoplasm, presumably due to weak binding of other metal ions that reach a higher concentration than calcium.

6
Limitations and Future Directions

While NMR spectroscopy provides a wealth of information about biological macromolecules ranging from structure to dynamics and binding events, its biggest limitation in general is its inherent low sensitivity. Because of this low sensitivity relatively high concentrations of the observed macromolecules are necessary. Since most proteins occur in the cellular environment at the low μM to nM level the effects caused by highly over-expressing a particular protein have to be carefully considered when designing and interpreting in-cell NMR experiments. In particular interaction studies between the overexpressed protein and potential interaction partners are difficult or even impossible unless the interaction partner reaches the same high intracellular concentrations as the observed protein. The current detection limit for [^{15}N,^{1}H]-HSQC-based in-cell NMR experiments is approximately 200 μM [7, 11], while the corresponding sensitivity for methyl group [^{13}C,^{1}H]-HSQC experiments is approximately 70 μM [13].

On the basis of these limitations, in-cell NMR experiments are most useful for the investigation of the folding state of proteins, the binding of small molecules that are highly abundant or are added externally or whenever a catalytic relationship between the interaction partner and the protein of interest exists, for example proteins that become phosphorylated or proteolytically processed. Furthermore, in-cell NMR experiments can provide information about the overall binding and interaction state of the protein in the cellular environment. While proteins like NmerA tumble freely in solution, others like thioredoxin interact with other cellular components, making them visible only in methyl group-based NMR experiments.

The final goal of in-cell NMR experiments is, however, the observation of proteins at or near their physiological concentration. For most proteins, this goal can only be achieved if the sensitivity of NMR spectrometers can be significantly improved. In recent years, the introduction of cryogenic probes has dramatically increased the sensitivity of NMR instruments, however, further significant improvements are necessary if the goal of observation of proteins under physiological conditions is to be met. Currently, the biggest challenge for in-cell NMR spectroscopy, however, is its extension to eukaryotic cells. Some preliminary experiments with yeast, insect cells and in particular with injection of proteins labeled with NMR-active isotopes into Xenopus oocytes [21, 22] have shown that experiments with eukaryotic cells are in principle possible. For sensitive eukaryotic cells such as insect cells or even mammalian cells further improvements of the quality of in-cell NMR experiments and a concomitant reduction in the required over-expression level have to be achieved. One particularly important factor for these cells is the survival time under NMR-conditions. While classical in vivo NMR experiments with attached cells have been performed in special NMR tubes cells that are capable of surviving in suspension are more suitable for in-cell NMR experiments since cellular suspensions can reach the required cellular density that is necessary to obtain the desired signal-to-noise-ratio. Suspension cells, however, pose another technical difficulty. To ensure their survival (and metabolic stability) for a time period long enough to cover the NMR measurement time they have to be supplied with oxygenated and nutrient-rich media. Fortunately, modified NMR tubes and gel entrapment techniques that keep the cells inside the tube but at the same time allow for an exchange of media have been designed and used in classical in vivo NMR investigations [36, 37]. Such devices and procedures will allow researchers to extend the measurement time of the experiments which will further decrease the detection limit. Initial experiments with agarose-trapped bacteria expressing NmerA have been successful [15] and indicate that in-cell NMR experiments with more sensitive eukaryotic cells might be possible in the future.

References

1. Wüthrich K (1989) Science 243:45
2. Foster MP, Wuttke DS, Clemens KR, Jahnke W, Radhakrishnan I, Tennant L, Reymond M, Chung J, Wright PE (1998) J Biomol NMR 12:51

3. Fesik SW (1993) J Biomol NMR 3:261
4. Shuker SB, Hajduk PJ, Meadows RP, Fesik SW (1996) Science 274:1531
5. Reese ML, Dötsch V (2003) J Am Chem Soc 125:14251
6. Serber Z, Keatinge-Clay AT, Ledwidge R, Kelly AE, Miller SM, Dötsch V (2001) J Am Chem Soc 123:2446
7. Serber Z, Ledwidge R, Miller SM, Dötsch V (2001) J Am Chem Soc 123:8895
8. Serber Z, Dötsch V (2001) Biochemistry 40:14317
9. Wieruszeski JM, Bohin A, Bohin JP, Lippens G (2001) J Magn Reson 151:118
10. Dedmon MM, Patel CN, Young GB, Pielak GJ (2002) Proc Natl Acad Sci USA 99:12681
11. Hubbard JA, MacLachlan LK, King GW, Jones JJ, Fosberry AP (2003) Mol Microbiol 49:1191
12. Shimba N, Serber Z, Ledwidge R, Miller SM, Craik C, Dötsch V (2003) Biochemistry 42:9227
13. Serber Z, Straub W, Corsini L, Nomura AM, Shimba N, Craik CS, Ortiz de Montellano P, Dötsch V (2004) J Am Chem Soc 126:7119
14. Serber Z, Corsini L, Durst F, Dötsch V (2005) Methods Enzymol 394:17
15. Reckel S, Löhr F, Dötsch V (2005) Chem Bio Chem 6:1601
16. Reardon PN, Spicer LD (2005) J Am Chem Soc 127:10848
17. Bryant JE, Lecomte JT, Lee AL, Young GB, Pielak GJ (2005) Biochemistry 44:9275
18. McNulty BC et al (2006) Protein Sci 15:602
19. McNulty BC, Young GB, Pielak GJ (2006) J Mol Biol 355:893
20. Burz DS, Dutta K, Cowburn D, Shekhtman A (2006) Nat Methods 3:91
21. Selenko P, Wagner G (2004) In: 45th ENC, Asilomar, Pacific Grove, April 10–15, 2004
22. Serber Z, Liu C, Dötsch V, Ferrell JE (2004) In: 45th ENC, Asilomar, Pacific Grove, April 10–15, 2004
23. Goto NK, Gardner KH, Mueller GA, Willis RC, Kay LE (1999) J Biomol NMR 13:369
24. McIntosh LP, Dahlquist FW (1990) Q Rev Biophys 23:1
25. Waugh DS (1996) J Biomol NMR 8:184
26. Chin JW, Cropp TA, Anderson JC, Mukherji M, Zhang ZW, Schultz PG (2003) Science 301:964
27. Deiters A, Geierstanger BH, Schultz PG (2005) Chem Bio Chem 6:55
28. Sippel A, Hartmann G (1968) Biochim Biophys Acta 157:218
29. Campbell EA, Korzheva N, Mustaev A, Murakami K, Nair S, Goldfarb A, Darst SA (2001) Cell 104:901
30. Suzuki M, Zhang J, Liu M, Woychik N, Inouye M (2005) Mol Cell 18:253
31. Venters RA, Coggins BE, Kojetin D, Cavanagh J, Zhou P (2005) J Am Chem Soc 127:8785
32. Liu G, Aramini J, Atreya HS, Eletsky A, Xiao R, Acton T, Ma L, Montelione GT, Szyperski T (2005) J Biomol NMR 32:261
33. Kay LE, Torchia DA (1991) J Magn Reson 95:536
34. Pellecchia M, Meininger D, Dong Q, Chang E, Jack R, Sem DS (2002) J Biomol NMR 22:165
35. Hajduk PJ, Augeri DJ, Mack J, Mendoza R, Yang J, Betz SF, Fesik SW (2000) J Am Chem Soc 122:7898
36. McGovern KA (1994) In: Gillies RJ (ed) NMR in Physiology and Bionedicine. Academic Press, San Diego, p 279
37. Egan W, Barile M, Rottem S (1986) FEBS Lett 204:373

Author Index Volumes 251–283

The volume numbers are printed in italics

Subject Index

MIX
Papier aus verantwortungsvollen Quellen
Paper from responsible sources
FSC® C105338

If you have any concerns about our products,
you can contact us on
ProductSafety@springernature.com

In case Publisher is established outside the EU,
the EU authorized representative is:
Springer Nature Customer Service Center GmbH
Europaplatz 3, 69115 Heidelberg, Germany

Printed by Libri Plureos GmbH
in Hamburg, Germany